Adam Phillips

ATTENTION SEEKING

Adam Phillips, formerly a principal child psychotherapist at Charing Cross Hospital, London, is a practicing psychoanalyst and a visiting professor in the English department at the University of York. He is the author of numerous works of psychoanalysis and literary criticism, including *On Getting Better*, *On Wanting to Change*, *In Writing*, *Unforbidden Pleasures*, and *Missing Out*. He is the general editor of the Penguin Modern Classics Freud translations and a Fellow of the Royal Society of Literature.

ATTENTION SEEKING

Adam Phillips

PICADOR

New York

Picador
120 Broadway, New York 10271

Owing to limitations of space, all acknowledgments for permission
to reprint previously published material can be found on pages 121–123.

Library of Congress Cataloging-in-Publication Data
Names: Phillips, Adam, 1954– author.
Title: Attention Seeking / Adam Phillips.
Description: First American edition. | New York : Picador, 2022. |
   "Originally published in 2019 by Penguin Books, Great Britain"—
   Title page verso.
Identifiers: LCCN 2021037920 | ISBN 9780374539276 (paperback)
Subjects: LCSH: Attention-seeking.
Classification: LCC BF637.A77 P45 2022 | DDC 153.1/532—dc23
LC record available at https://lccn.loc.gov/2021037920

Our books may be purchased in bulk for promotional,
educational, or business use. Please contact your local bookseller
or the Macmillan Corporate and Premium Sales Department at
1-800-221-7945, extension 5442, or by email at
MacmillanSpecialMarkets@macmillan.com.

For book club information, please visit facebook.com/picadorbookclub
or email marketing@picadorusa.com.

picadorusa.com • instagram.com/picador
twitter.com/picadorusa • facebook.com/picadorusa

1   3   5   7   9   10   8   6   4   2

For Marianne

What we mean by feeling 'interest', I think, is the free acceptance of the gift of pleasure.

Robert Pinsky, 'Poetry and Pleasure'

I needed the sort of accomplices who were either completely faithful to the idea that what I was doing was legitimate, or else were not entirely aware of what my plan consisted of.

Muriel Spark, *Loitering with Intent*

. . . the romance of motifs without the banality of motives.

Matthew Bevis, 'My Coincidences'

Self-consciousness anticipates an excess of seeing.

Jenny Xie, 'Visual Orders'

Contents

Attention-Seeking

I

It is awfully important to know what is and what is not your business.

Gertrude Stein, 'What is English Literature'

Everything depends on what, if anything, we find interesting – on what we are encouraged and educated to find interesting, and what we find ourselves being interested in despite ourselves. And when we are interested, we pay attention: sometimes, at considerable cost. There is our official curiosity and our unofficial curiosity: our official curiosity is a form of obedience, an indebtedness to the authorities. In our unofficial curiosity we don't know who we want to be judged by. It is the difference between knowing what we are doing, and following our eyes.

It is through both kinds of interest that we tend to recognize and characterize ourselves and other people. 'We get hooked,' the critic Aaron Schuster writes in *The Trouble with Pleasure*, on certain things, impressions, patterns, rhythms, words that give a warped consistency to our world, the grain of madness that provides us with our style and character, our secret coherence – whether this saves us or drives us to our doom.

Our attention is attracted, and we attract attention, in very specific, idiosyncratic ways. We are not hooked by anything and everything; we don't desire everybody; only particular people, images, things, patterns, rhythms and words affect us. Indeed, what is striking, as Schuster suggests, is just how selective we are, and how much we assume our coherence, however secretly; as though everything about ourselves could be connected if only we had the wherewithal. Whether or not it is a grain of madness that provides us with our style and character, this assembling of our selves through what we notice, through what, as we say, attracts our attention – both consciously and unconsciously – and just how surely we limit the repertoire of what we do notice, smacks of addiction ('we get hooked'); and of a fundamental unknowingness about how we make ourselves up. As though what we call our identity, which is to do with what we notice, is a kind of fixation, an obsession with certain ideas about ourselves. What we might call our taste, or more simply our preferences, becomes a type of fate, or a preferred picture of ourselves (which 'saves us or drives us to our doom'). The famous surrealist motto, 'Tell me what you are haunted by and I will tell you who you are,' all too easily translates into 'Tell me what you are interested in and I will tell you who you are.'

There is an assumption in psychoanalysis, as in the wider culture, that we are by nature interested creatures, driven to pay attention (at least once we have learned what it means to pay). That growing-up, ideally, means discovering one's interests; initially our apparently innate interest in our own survival, and our imaginative elaboration of this; and then, depending on our affluence and our inventiveness, our following of our curiosity as far as we are able. We may think of our attention as inspired by need, and formed by nurture (attention as another word for appetite). Babies and young children are, as we know, very intent on what they want, very intolerant of frustration, and very troubled by being bored, by losing interest in things. So we are prone, as adults, if we are lucky enough, to take our interests for granted, rather than be unduly bothered by them. Only when they become in some way disturbing do we become interested in our interest ('interest' then being a word for phobia, or obsession, or perversion, or addiction, or ideology, or hobby, or discipline). Our so-called symptoms narrow our minds by forcing our attention; and reveal, by the same token, just how much it is forced attention that we suffer from (so-called sexual perversions confine our interest in sex; anxiety and depression over-focus our attention; and this may be part of their function). It is, though, one of our projects to circumscribe the range and intensity of our curiosity; as though our capacity for interest was itself threatening, by being so potentially promiscuous, or unbounded, or unpredictable. As though we always have too much or too little appetite, too much or too little danger. Our interest in anything or anyone threatens to become excessive, or not excessive enough.

Whatever or whoever it is that does interest us, like the appetites that prompt our interest, effectively organizes our lives for us; we follow, and/or avoid following, our attention. Our interests are what we do, who we listen to, where we go. Psychoanalysis redescribes interest and attention as sexual desire, and takes this to be the informing force and purpose of our lives; and clearly sexuality can be used, if only by analogy, as a way of seeing how interest and attention might work (any presumed life force, or essence, is always an explanation of attention and interest). So when one of the early psychoanalysts, Ernest Jones, introduced the Greek term 'aphinisis' (disappearance) into psychoanalysis, he was addressing one of our fundamental terrors: the 'primal anxiety' of having no interest, 'losing the capacity or opportunity for obtaining erotic gratification', loss of desire as loss of life; a predicament, in other words, in which satisfaction is impossible: in which nothing is of interest, nothing engages us, and we are not drawn to anyone (it is perhaps akin to the nineteenth-century fear of the sun going out). Contrasting it with what he calls 'the artificial aphinisis of inhibition', real aphinisis is the 'total extinction' of 'sexual capacity and enjoyment as a whole'. Desire might feel or have been made to feel so unbearably conflictual that it has to be abolished; a person is then left living in a world in which there are, to all intents and purposes, no objects of desire. What can we do with ourselves when, or if, nothing is of interest? And what would we have to have done, or what would have had to happen to us, for our interest in life to disappear?

We need to wonder, then, why we would ever want to accuse anyone of being attention-seeking. Attention-seeking is one of the best things we do, even when we have the worst ways of doing it. In its familiar sense, it is a way of wanting something without always knowing what that is. And it is, by the same token, a form of sociability, an appeal to others to help us with our wanting. Whether it is a cover story for the straightforwardness of desire, or a performance of the perplexity of demand, we seek attention without quite understanding what the attention is that we seek, and what it is in ourselves we need attending to. It is out of this complexity that people get together, to find out what is possible (sociability depends on attention-seeking). Part of the apparent relief of acquiring language is that it seems, occasionally, to clarify the obscure exchanges we make our lives out of.

We are attention-seeking, in both senses – throughout our lives, and not only as children – partly, as I say, because it is not always clear what we are seeking

7

attention for, and what we want to pay attention to: what in ourselves, and outside ourselves, needs attending to, and what we hope will be the consequences of securing the attention we seek. Because attention-seeking is not generally prized, it has to be disguised as something else – as art, say, or manners, or prayer, or success – so a lot of our so-called creativity involves us in finding acceptable ways of finding and attracting the attention we desire. This attention may misfire, may not get us the lives we want, but attention-seeking is where we start, and what we start with. What Lee Edelman calls, with deliberate tastelessness, 'the fascist face of the baby' is drawing our attention to our earliest gifts and talents for getting people to notice – or rather, to reveal – what we need; the attention we have to give to being paid attention, and the attention we have to give to what is worth paying attention to.

Need requires attention, and everything relies, initially, upon the kind of attention that meets our needs; and then on the kind of attention we can give to our needs and wants as we grow up (needs are constituted by the ways in which they are responded to). Everything follows on from how and where we pay our attention; both the attention that is demanded of us, and the attention we give without intending to, without noticing. The bringing-up and educating of children, whatever their culture or class, initiates them into regimes of attention; it tells them, in no uncertain terms, what is worthy of their attention, and how it should be paid, as well as what kind of attention they should be wanting, and how they should go about getting it (neither distraction nor showing off is taught in schools). All religions, moralities, arts, sciences, politics and therapies organize and promote certain kinds of attention; in their different ways they tell us where to look and who to listen to; they tell us what about ourselves we should value and be valued for: what about ourselves we should take an interest in, and what we should take rather less interest in than we do.

And yet, of course, no one can actually predict the consequences of the attention they pay and are paid. As education and propaganda and advertising and sexuality continually reveal, people's attention can be exploited and manipulated and directed, but it can't be ultimately controlled. We never quite know what people will make of what they are given; or how their minds may drift while they are paying attention. These essays, then, are about this curious and revealing phrase, 'paying attention'; not quite a cost–benefit analysis, but to do with investments and risks, and questions about currency and exchange rates. By attention, I mean simply how we find and involve ourselves with what interests us, what encourages and what inhibits us in following our curiosity, and what effect our being interested may have on ourselves and other people. As these essays want to suggest, it is worth noticing both what provokes our attention and what, like shame, inhibits it. So they are also about what the philosopher Gilles Deleuze calls our 'capacity to be affected'.

We have to back our interests to find out what they are; and ideally our earliest environments should allow for and encourage certain of our interests – what John Stuart Mill called 'experiments in living' – rather than simply or solely determining what they should be (morality not born of experimentation can only be dogma). Attention and interest are always themselves experimental, even when – or perhaps particularly when – we are unaware of the risks being taken; curiosity never comes with a guarantee. Clearly, to prefer safety to curiosity, or to experience them as too much at odds with each other, is to limit the possibilities of experience, as is a consistent or too certain knowing of what one is interested in (it is always worth wondering what our interests are a way of not being interested in). Having been, at first and of necessity, consistently interested in and attentive to our parents, it is always an extraordinary moment in a child's life when she begins to realize that there are pleasures outside the family, that her parents' words are not the only words in the world (one of the ways we look after our parents is by believing what they say). One of the child's fundamental questions is: what is he allowed to be interested in outside the family? The history of our attention, in other words, is one of the stories of our lives.

II

Language only ever shows you how things would look
if language was used.

Miles Hollingworth, *Ludwig Wittgenstein*

There is, of course, a great deal of concern currently
about the effect of modern technologies on attention,
and particularly the attention spans of the young (attention spans, to some people's relief, unlike qualities of
attention, can be measured). In one of many clearly heart-
felt contemporary jeremiads, the philosopher Talbot
Brewer writes in a striking recent essay, appropriately
entitled 'What Good are the Humanities?':

> Our attentional environment has not equipped students
> with the traits required for appreciative engagement
> with literature or art or philosophy; the habits of
> devoted attention and of patience and generosity in
> interpretation; the openness to finding camaraderie
> and illumination from others in the more treacherous

passages of life . . . What degrades our attentional habits is not just advertising itself but also those communicative forms that survive the fierce competition for advertising dollars.

Of course, many more people are interested in those 'communicative forms' – social media – than in literature, the arts and philosophy; and not everyone values devoted attention, patience and generosity – which do not necessarily go together – nor, indeed, the acquiring of 'good' habits (it is part of the consensus being assumed here that 'we' all know what these words mean, and why they matter; what is being promoted here is the right kind of attention to what words are used to do). But it is not unusual now for loss of attention to be equated with loss of morality, if not the loss of culture itself (that is, the culture defined by those people who claim to be in a position to decide what culture is). Nor is it unusual to assume that changes in habits of attention portend larger changes; and that our morality or civility is somehow bound up in the ways we pay attention: in what we make of our attention, and what we make out of it.

I was educated in the culture that is being defended here – a culture of close reading and slow looking, in which the arts were taken to be formative – which tended to pride itself, as certain kinds of liberalism do, on the ability and freedom to question and cast doubt upon itself; and to concentrate in a way that made recognition of something other than one's self and its abiding preoccupations possible. As though this, in and of itself, guaranteed something; something that could be called the impossibility of fanaticism, or fascism, or certain kinds of egotism; or the imagining of lives not devoted to profiteering; or the possibility of including as many people as possible, in as many ways as possible, with as little demeaning hierarchy as possible (all in the service of personal development, assumed to be the ultimate value). Which at its best it can do and at its worst is as coercive and sure of itself as the people it claims to oppose. Liberalism always claims to be widening our attention (and sympathy) without always being able to know what to do with the attention it has made possible (and, indeed, whether we can bear the sympathy we find ourselves feeling). We have to wonder what the good reasons are for regulating people's attention, and what the best ways of doing so are; and also what we imagine unregulated attention would be like (undevoted attention is presumably attention devoted elsewhere). The catastrophe is always of people being too interested in the wrong things, in the wrong ways; for the fascist and the antifascist alike, for example, it is by definition very difficult to give attention to anything other than fascism. Morality is always about seeing things as they should be seen: the right kind of attention pointed in the right direction, the right language about the right forms of attention.

Perhaps unsurprisingly, psychoanalysts of very different schools have taken up this issue of the widening of attention. Following on from Freud's idea of a new way of listening, which he called 'free-floating attention' – a kind of attention in which one doesn't know beforehand what is of interest – Christopher Bollas (of the Independent Group) has written of people's fear of the complexity of their own minds; Marion Milner has written of 'wide-angled attention', which is receptive by not focusing (when I paint a tree in a field, she once told me, I look at everything except the tree); Michael Feldman (a Kleinian) has written of the active narrowing of the mind that is required for people to attack their own development; and the French analyst Jacques Lacan remarked that the psychoanalyst starts explaining things to the patient when he, the analyst, begins to be frightened of his own curiosity. Each of these writers is alerting us to anxieties about attention and curiosity, while not always being able to give us an account of what a person's life might be like if they were unfrightened of their own complexity, or refused to determinedly narrow their minds, or were unintimidated by their unbidden curiosity. What, after all, is it imagined could happen between an analyst and her patient – or indeed any couple – if they were less frightened of their curiosity, less focused on what they imagined they wanted from each other? If explanation is the self-cure for curiosity, we have a lot of explaining to stop doing; if desire is the refuge from wide-angled attention, we have a lot of wanting to relinquish. And yet here, as everywhere in such discussions, the wrong kinds of attention are shown to make us suffer, but the consequences of the right kinds of attention are somehow unelaborated, shied away from. What would a complex or a broader mind be good for? What would a more curious life be like to live? What kind of sociability, or sexuality, would it entail? What would our lives be like without focus? Or without our attention always being prescribed, or described as a means to a known end?

I started what turned out to be this book after writing a talk – not incidentally in the light of Talbot Brewer's account of the so-called humanities – about the literary critic Stephen Greenblatt (included in the Appendix). It was striking to me that Greenblatt was preoccupied by distraction – language being one way of paying attention – and that, unsurprisingly, distraction was a way in to thinking about the vagaries of attention (and, of course, with how reading itself confronts us with questions about attention, and about the quality and purpose of our interests). Both psychoanalysis, as a writing and a therapeutic practice, and the reading of literature – literature and psychoanalysis being the informing presences in this book – work by making us self-conscious about the nature and the quality of our attention, our language; by drawing our attention to certain preoccupations, they make us wonder what our attention may be seeking and avoiding. Most people are not, and never will be, interested in 'literature', or literary criticism, or indeed in close reading; but everyone's

attention is absorbed by something, even if it is only distractedness, or the lack of control they have over their
attention. And many people seek out and value experiences in which they can lose themselves, or become
absorbed (psychoanalysis, at its best, contrary to popular
prejudice, is the therapy that frees people to lose interest
in themselves; there's nothing more self-preoccupying
than a symptom, nothing finally less interesting than
one's self). The kinds of interest we take, the forms of
attention we prefer, seem to be the best ways we have so
far of trying to get the lives we think we want.

To begin with, the question is always: at any given
moment, what is worth paying attention to? And then,
what kind of attention should we be paying? And
then, what are the reasons we can give for doing this?
And then, how do we decide between reasons? And
then, what are our criteria, and who do they come
from? We must be interested in the right things in the
right way. Or at least this is what everybody tells us.

And yet, in Freud's description of dreaming, attention becomes the word for a multiplicity of competing
aims and wants. Freud gives us an account of a new
form of attention, unknowing attention, in which we
take in the world around us in ways, and for reasons,
that we are unaware of. Dreaming gives us a picture of
how we are interested otherwise in our lives. And
whether or not we agree with Freud's account, it shows
us what it might be to pay our lives a different kind of

attention; and, moreover, to pay our attention a new kind of attention. During what Freud calls 'the dream-day', for example – the day before the night of the dream – we are not paying attention in all those ways we are supposed to, or imagine that we are supposed to. We are, Freud suggests – attentive as ever to our possible future satisfactions – gathering material for what he refers to as 'the dream-work' we will do when we are asleep, but without registering that this is what we are doing. During the day we are like sleep-walkers (and workers) preparing for our dream. During the dream-day, then, we are unwittingly at work on that night's dream. And this involves our being interested not in the right things – the taken-for-granted obvious objects of interest – but in unpredicted, quite unexpected things; and being interested in the unpredicted things in an unpredictable way. Freud's concept of dream-work, in other words, radically revises our notions of the attention we thought we could pay. Dream-work shows us that we are more interested, and differently interested, in the things that interest us, and also in those in which we seem to have no interest at all. And that we are all the time unconsciously paying attention in ways we are, by definition, unaware of, to things we didn't know mattered to us. We are more than interested in the day because we never know beforehand what it can provide, or what it contains (and we can never know beforehand, never predict, the dream we will have at night). Whatever else we are, we are the feral children of our day, but alert for what we didn't know we needed; mindful of the dream ahead.

As we know from experience, things crop up in our dreams – images, phrases, gestures – that we have either forgotten we noticed during the day, or that we were absolutely unaware of noticing. Freud writes in *The Interpretation of Dreams* about 'the exclusiveness of the claims of the day immediately preceding the dream', suggesting that the day makes claims on us, and we make claims on the day that we are unaware of; and that these claims that always go unnoticed have some kind of urgent priority, an 'exclusiveness'; and, indeed, that our days are there for us to make dreams out of (to make a dream, in Freud's view, is to formulate unconscious desire; and that is the way we go on making a future, working out what we want and need). It is as though there were a figure inside us – call it 'the dream-maker' – who is interested in our day on our behalf, but without our ever realizing what is of interest, what might be of use, in that day. The dream-work – the work of whatever about us makes our dreams – is, Freud writes, 'under some kind of necessity to combine all the sources which have acted as stimuli to the dream into a single unity in the dream itself'. And in a footnote he writes of 'The tendency of the dream-work to fuse into a single action all events of interest which occur' in the dream-day. It is not surprising that many people have noticed that the way Freud describes the making of a dream sounds similar to the making of a work of art, the dream-maker collecting the materials for the dream during the day, but without telling us; or indeed telling us that there is a dream-maker at work (Nietzsche had remarked, as Freud must have known, that 'we are all artists in our dreams').

The invisible artists of our own desires, we are in Freud's account astoundingly attentive to our making and our medium. And the made thing, the dream – like some works of art – tends to be intriguing but without being obviously revealing; all our unconscious attention leading us to something we don't quite know how to pay attention to (if there was a key to understanding dreams, or indeed art, we would all be using it; if, that is, understanding is what we want. Freud sometimes intimates that in the wish to understand we can be paying the wrong kind of attention). People have always interpreted dreams without there ever having been *the* interpretation of dreams; people have always worried away about what kind of attention they should give to their dreams, and about what kind of attention might have gone into the making and the dreaming of dreams (in the making of a dream, we pay attention in the service, paradoxically, of making something enigmatic that can baffle and excite our curiosity). Freud was as interested in the nature of our attention – how we use it and what we use it for, what kind of medium it is – as he was in sexuality. Indeed sexuality in his work is often there to explain the conundrums of attention.

The psychoanalyst Charles Rycroft writes in *The Innocence of Dreams* – the title itself suggesting the sense in which we are innocent of our making, and so of our initiating attention – that 'Dream formation requires an instigator or trigger-event in the dreamer's waking life – this instigator bearing some actual or symbolic relation to a repressed impulse striving for expression.' The instigator is whatever reminds us of something about our unconscious desires. But we should note that we know neither what the trigger is when it occurs, nor, consciously, what our forbidden desires are that are being triggered. Something about us, as it were, knows a trigger event when it sees one. In this story we are leading a double life: a life of conscious interests and concerns, and a life of as yet unconscious desires. We may be being encouraged, Freud intimates, to pay attention to the right things in the right way, but there is another side of ourselves that is working on something quite different – what else we might want, what else is of interest to us – and that will be made into a dream (in this sense, dreams are the counter-culture of the culture). I may go to a talk and be very interested and preoccupied by what is said; but that night I might dream about the speaker's tie. And were I then to give my associations in a psychoanalysis to the tie in the context of the dream – and of course in the context of my past and present life – I would most likely discover thoughts and feelings and memories quite different from anything said in the talk that so absorbed me. My interests and my attention, in other words, were not only, or simply, what I had taken them to be. There was far more in that talk, for me, than I had realized (the tie, as also a pun); it had been used by a side of me for what Freud called 'day-residues' – whatever I could use from the day to formulate my unconscious wishes in a dream. As though the day is used by the dreamer to find ways of formulating, through the dream, unconscious, unacceptable, as yet unrecognized desires. Or, to put it another way, the day is being used to make out a possible future, and to make a future out of. 'The clearest sign of a truly creative, self-renewing mind,' the critic Geoffrey Hartman wrote, 'is to build up greatest things from least suggestions'. We are all, in Freud's view, truly creative and self-renewing as dreamers.

Whether or not one accepts Freud's explanation of the content and function of dreams set out in *The Interpretation of Dreams* – of which this is a cursory account – it is clear what he is trying to make out of his theory. Freud's theory of dreams is a story about our unnoticed noticings, about our unofficial curiosity, about the interests we have, and the kinds of attention we give that we don't, or would rather not, pay attention to, or when we are, to all intents and purposes, not paying attention at all. What happens, he is asking, to the attention we give to whatever is not socially sanctioned, or endorsed, or even just preferred by ourselves? He is drawing our attention not merely to the all too familiar but never really familiar sense we have that there are desires and intentions and ambitions we would prefer not to acknowledge, but also to the sheer scale of our interests, to the exorbitant unpredictability of our attention; to the ways we obscure our possibilities. Calling it sexuality may already be to circumscribe, to over-code, what is being referred to.

What does the child do when she finds herself with a sense of unheard-of possibility? When she keeps noticing and enjoying what is either unaccounted for or actually disapproved of in her early environment, something her language doesn't cover? It is clearly a crucial moment in a child's life when he begins to notice that there are pleasures outside the family, when he has seen that there is something he wants outside the family circle. We should assume that the vast range of our unofficial, unapproved-of interests may have to go on taking stranger and stranger forms in their baffled quest for recognition (so-called symptoms encode the as yet unregistered experience of desire). That we are always at work on, and trying to avert the loss of, what Lacan refers to as our 'missed encounters'; missed partly because we were unable to find a way of articulating our obscure glimpses of what we might be wanting; or missed because they have been shied away from, or because they were not recognized as encounters. It is here, of course, that the culture rushes in to tell us what we must have been wanting and seeing, and so does not give us the space – the kind of education – to find out. To be interested in attention is to be interested in the work, the inventiveness of wanting, and the perception it founds and informs. And to acknowledge that what we might want may not as yet exist. That wanting necessitates making.

So I think Freud was wrong to assume, in *Civilization and Its Discontents* (and elsewhere), that 'in mental life nothing that has once been formed can perish – that everything is somehow preserved' (when Freud used the city of Rome as his example of the imperishability of the past, he did so in the full knowledge that most of Rome had in fact disappeared). It might be truer to say – and Freud himself may have been intimating this in his commitment to a life apparently without loss – that many of our unnoticed noticings, our unofficial forms of attention, are preserved. They haunt us like ghosts that stay around awaiting their moment. What is lost is only of interest as a form of the possible (possibility is the only thing worth repressing). What we are looking for are opportunities with uncertain outcomes; and we are not actually in a hurry to discard them, as they occur. Unrealized opportunities, uncompleted actions, unfinished experiments are what we hold to, and hold on to. This is what our attention is for; when it is not being recruited, as opposition, to secure us against an unknowable future. There is attention in the service of reassurance, and attention in the service of something else.

Dreams, at least in Freud's account, become the first version, the first formulated stage of the possible (they unsettle rather than reassure: they are self-evidently the new); which, perhaps unsurprisingly, takes strange and initially unintelligible forms. Dreams, that is to say, are not in words; only reports of them are (they do not have the illusion of intelligibility that language always creates). They effortlessly challenge the attention we give them, while being all too often irresistibly interesting. And we can see them – as vividly immediate as they can be, and as quickly fading – but not with our eyes. In this sense they may have been the best emblem, for Freud, of the puzzles posed by our interest and attention. Dreams, he proposes, interest us in our interest; and his therapeutic method of inviting the patient to free-associate to elements of his dream reveals the dreamer as someone paying unexpected attention to his life. As though what we need to know something about is our attention; both the history of our forms of attention, and what we are using our attention to do. What we desire can distract us from how we desire; how we desire can distract us from what we desire. To be interested in anything, Freud suggests, we have to be sufficiently interested in our interest.

It is not news that we see what we want to see, that perception is distorted by wishes. It can be news that we see far more of what we want to see than we realize. This book is about how we go about doing and not doing this. About just how interested we are, whether we like it or not.

III

There is nothing deep down inside us except what we have put there ourselves.

Richard Rorty, *Consequences of Pragmatism*

If, as Freud and others suggest, we are also composed by, and composed of, our unnoticed noticings ('as the child grows', Winnicott writes, 'the content of his personal self is not only he'); if we are taking in and somehow registering much more than we are capable of being aware of, or indeed want to be aware of, this clearly has implications for, to use the title of one of Winnicott's most remarkable papers, morals and education. Winnicott, though, schematizes the growing child's extravagant receptiveness as a conflict between compliance and spontaneity (only the inhibited are spontaneous); the child's adaptation to his environment at the cost of his own personal development (the picture is of something innate and unfolding more or less distorted by the environment). Promoting what he calls, in 'Morals and Education', 'cultural phenomena lying around for the child to catch hold of and to adopt', he suggests that adults 'leave lying round not only objects (teddy bears, dolls, toy engines) but also moral codes' so that the child follows, according to his desire, what attracts his attention, rather than having a morality foisted upon him by the adults, which he can only comply with and copy, but not use in his own way (the picture here is of natural, innate, personal drifts of attention as opposed to coerced and manipulated attention). Thus the child 'grows his or her own moral capacity . . . and finds his or her own way of using or not using the moral code or general cultural endowment of the age'. If trauma is untransformable experience, then any moral belief – or, indeed, theory, psychoanalytic or otherwise – that is simply abided by rather than personally transformed is akin to a trauma. For Winnicott, the child's use of what is found lying around is his capacity to transform what he finds according to his own desire. So handed-down morality, any handed-down knowledge, is always potentially a dead end; and this of course radically revises traditional ideas about tradition. We can clearly translate Winnicott's picture of what parents and children can do together into later forms of acculturation, as the institution of the family joins up with, and is partly superseded by, the institutions of education.

There is clearly far more to teaching, for example (and following Winnicott's line), than what is nominally being taught. Teachers – not unlike some students – are by definition attention-seeking; but as with all attention-seekers, there can be a discrepancy between the attention they seek and the attention they get. And, indeed, a discrepancy between the attention they give and what the student receives (between what is presented and what is found lying around). By the same token, as it were, we may wonder what kind of attention we give – as opposed to the attention we are supposed to give – to morality. Or, to put it slightly differently, what kind of objects of desire are rules and regulations; if we take it that there is always more – or something else – to desiring than we can ever be fully aware of? My intention to follow a rule depends upon my myriad-minded attention. My attention to what is there lying around and my attention to what I am being told is important (Oscar Wilde was once asked at dinner, 'But Mr Wilde, don't you think morality is important?', to which he replied, 'Yes, but I don't think importance is').

If rules, like teachers, are by definition attention-seekers, they also try to define the attention they seek; as though we know what it is to be taught and to learn, just as we know what it is to follow a rule. Though it is quite clear that there is a difference between teaching *Hamlet* and teaching someone to play chess. The rules of chess are not up for discussion in the sense that *Hamlet* is; and yet in both, very different, examples the supposition is that appropriate forms of attention can themselves be taught, in order to make further teaching possible. The question becomes not only whether you can teach forms of attention – which clearly to some extent you can: people learn how to play chess and learn to read and watch and talk about *Hamlet*; it becomes: what do we think we are doing, from a psychoanalytic point of view, when we are teaching or learning forms of attention? What do the psychoanalytic presuppositions – of the unconscious, of sexuality, of what Freud calls the 'pain-and-pleasure principle' – add to or subtract from the conversation about attention-seeking, in both its senses: of wanting other people's attention, and of seeking satisfactory forms of our own attention? Can we tell the unconscious to pay attention? What kind of sense does it make to talk about educating the unconscious, or sexuality? Well, very little, but we seem to be doing it all the time. Or wanting to be doing it. Acculturation means getting your attention-seeking right.

In one sense we are talking about instrumentalizing Deleuze's notion of our 'capacity to be affected'; of being able to be pragmatic – practical and problem-solving – where pragmatism doesn't obviously work. If I am continually being affected by the world inside and around me, in ways that I am often unable to acknowledge, how educable am I, or what kind of education am I getting despite the intentions of the authorities? Do I, in Winnicott's language, always notice when I have picked something up that was just lying around (a song may stay in my mind)? Psychoanalysis is one way of describing what happens to us despite what we want to happen to us;

and a way of describing what we are doing despite what we think we are doing. When I make a so-called Freudian slip, I am seeking attention for something without realizing that that is what I am doing; and we can infer that I have been paying attention elsewhere. When I say something without apparently wanting to, my attention-seeking has gone awry; I have taken myself and other people in a different direction, an unintended direction. When I dream, I have taken myself to an unpredicted and unpredictable dramatic space, what Freud calls 'the other scene'. While dreaming, I pay involuntary attention; on waking, it is as though I can choose what kind of attention I can give to my dream. What Freud calls the interpretation of dreams involves, among other things, translating one form of attention, the 'dream experience', into another, the interpreting of the dream, through free association. In free association you see (and note) what occurs to you around and about the elements of the dream, the dream itself being something that can be said to have occurred to you during the night. In Freud's descriptions of what he refers to as 'psychic life', it is as though the dream and the slip are themselves seeking attention – wanting to be recognized for what they might be (representations of unconscious desire, in Freud's view); and that the dream and the slip are the ways we seek attention; they are the means by which we make enigmatic demands on ourselves and others. Enigmatic because we have to obscure them, because they are representations of unacceptable desire. And so like all forms of attention-seeking, we might assume that we don't always know beforehand what the attention is that we seek. And yet it is assumed in psychoanalysis – as it would be in any essentialist theory – that we do know what the attention is that we require; we know, unconsciously, what in ourselves needs attending to, and the desired forms of attention that we seek (after we have answered the question: what is a need? we have to ask the question: what is it to meet a need?). Needs are defined by the kind of attention we give them. And if anything is attention-seeking, it is a need.

So when Freud – and much of the psychoanalysis that follows on from him – prescribes the nature of our need, he is, we might say, helping us with our attention-seeking. But to essentialize need is to coerce attention. Once you have defined the nature of need, you have limited the repertoire of forms of attention; and over-defined the quest romance of attention-seeking by explaining its aims and objectives. So-called attention-seeking behaviour ceases to be an experiment in living; it becomes a project rather than a probe, a programme rather than a form of curiosity. So when Winnicott writes of culture as 'the imaginative elaboration of physical function', he is imaginatively elaborating the always potentially reductive Freudian accounts of culture as sublimated instinct; of instinctual drives seeking the gratifying attention called satisfaction, and of that being the be-all and end-all. It depends on where and why you want to put the stress, on the physical function or the imaginative elaboration. After all, once you begin on the redescription of physical function that Winnicott calls imaginative elaboration, where will it end? What will physical function seem like in the light of its imaginative redescription? In essentialist theories you can only end up where you began (with God, original sin, instincts, genes, etc.). So I want to suggest that attention-seeking may be very unlike our other seekings. That to seek attention is to seek out the nature of our seeking; and at its best it is a means (and a medium) with no foreseeable end (we could say now that the seeking of attention is the seeking for something that is not a commodity). Attention-seeking, whatever else it is, is always a love test, and should be treated as such. That is, without contempt. In our attention-seeking it could be assumed that we know neither what we want nor what we expect; and so we are in our starkest dependence on others. And in that true state of absolute dependence lies the possibility, the groundswell, of new forms of sociability. Attention-seeking then, ideally, as a comedy of errors, rather than a tragedy of failures. Attention-seeking as something that might come without excuses.

The critic Eric Griffiths makes the best case for this in writing about comedy in relation to J. L. Austin's great essay 'A Plea for Excuses'.

Remembering Austin's philosophically grounded plea for a keener attention to a greater variety of examples, it will, I hope, be clear that it is not in a spirit of anti-intellectualism, or anti-theory that I offer no formulation of the essence of comedy. Thinking about comedy does have to submit to the test and the pleasure of examples, and has to deny itself the relaxations of the Big Idea. So I end not with a conclusion . . .

Keener attention to a greater variety of examples of so-called attention-seeking will be sufficiently interesting and enjoyable only when and if we can relinquish the need for what Griffiths calls 'the relaxations of the Big Idea' (so attention might be a more useful idea than human nature). Psychoanalysis at its best is a way of letting a greater variety of examples have their say; of enjoying the test and the pleasure of examples without foreclosing on them. Rather than exposing the relaxation of the Big Idea – of seeing the Big Idea as the problem rather than the solution – Freud was always tempted to present psychoanalysis as the Biggest Big Idea. He opened something up – a world, one might say, of various examples and uncontainable meanings – and then closed it down with a few Big Ideas. The attention-seeking that psychoanalysis has always done – and the attention-seeking it uses and redescribes in its treatment – has all too often been in the service of the Big Idea. And the Big Idea always requires a restricted and restrictive attention. The Big Idea now may be that a Big Idea is no longer needed.

Shame and Attention

I

Intelligence does not really exercise *free choice*: it can only *select* among the offerings the system affords. Intelligence follows the logic of a system.

Byung-Chul Han, *Psychopolitics*

When the critic D. A. Miller writes that what 'lies at the close heart' of Jane Austen's style is 'a failed, or refused, but in any case shameful relation to the conjugal imperative', he invites us to imagine what a shameless relation to marriage might be. Perhaps Mickey Rooney's suggestion that he always got married in the morning so if it didn't work out he wouldn't have wasted the day? It is, that is to say, always worth wondering what any apparently shameful act would be like if done shamelessly (as if we could then work out what we do to our experience by adding or subtracting shame: as though we could choose to feel it). A shameful relation to marriage – or to the 'conjugal imperative', as Miller puts it, clearly relishing both terms – might be a complicated thing. Indeed, the ambiguity of the phrase could suggest that one might have a shameful relation to being married, and to not being married, despite the fact that we know Austen was not herself married. And, of course, there are gay men – to whom Miller is implicitly alluding, without quite making Austen a queer theorist – who think that gay men should be ashamed of themselves for wanting gay marriage, for wanting to make that concession to straight life, with all it entails. As though marriage is also something one should have a shameful relation to. A shameful relation to anything is, by definition, a determined narrowing of attention. We can only pay shame a certain kind of attention. And we can only pay ourselves a certain kind of attention when we are mortified.

By referring to a shameful relation to marriage, Miller both ups the stakes and intimates that all shameful relations are not straightforward, or straightforwardly bad. Austen's shameful relation to marriage, after all, is integrally related to her much-acclaimed style; it is, in Miller's account, among the sources of something that is one of the best things about her. No one says shame is a blessing (or a gift, or a talent); but we might wonder, in the pragmatic way, what we might be using shame to do – and to do to our attention. Or how shame might be used in, say, Austen's case to conjure up a style, or a morality, or a speculation about what shame might make possible. And this seems particularly improbable because shame is virtually by definition one of those feelings – one of those experiences – that might seem, as people used to say, irreducible. So unbearable, as it so often is, as to be unusually difficult to think about. Though when we are thinking about it, we should probably bear in mind that the *OED* suggests that 'shame' and 'sham' may be linked etymologically: 'sham' is 'commonly explained as in some way connected with "sham", northern dialect form of shame'. When we are at our most authentic, it may be worth wondering what we may be pretending to.

'When you are doing anything concrete,' Ford Madox Ford once remarked, 'your powers of observation desert you.' There is nothing less abstract, more visceral, than feeling ashamed; and so, by the same token as Ford suggests, your powers of observation desert you. In an uncanny narrowing of the mind, shame forecloses one's attention. And this, as we shall see, may be part of its function. One of the projects of shame is to create blind spots. So shame is a certain kind of attention, a quality of attention (or inattention) that one gives to the self (and an emblem, by the same token, of what we are capable of doing to, and with, our attention). One might wonder, as of any form of attention, what is wanted by giving this appalled attention to oneself in which it is as though one part of the self rises up, as it were, and assumes a disgusted, omniscient superiority over what seems like the entire self. And in which one might also, indeed, legitimately wonder – given the character assassination that is shame – which is the more horrifying, the judge or the judged. Though this is a question that is unlikely to be asked, because when one is ashamed of oneself, it is generally presumed that one is thoroughly deserving of one's mortification. For shame to have its effect, there must be no mental space for second or third thoughts about what has happened and is consequently happening. There is simply the torturer and the tortured. Shame is a figure for the mind colonized.

'What arouses shame,' the philosopher Bernard Williams writes in *Shame and Necessity,* yoking the two terms together, as though both were ineluctable, 'is something that typically elicits from others contempt, or derision, or avoidance.' And that typically elicits something even worse from the self. Shame, that is to say, is a conversation-stopper. Once someone's actions are deemed beyond justification, the limits of exchange have been reached. It is a question that often confronts both the priest and the psychoanalyst: what can you say to or for the person who is ashamed? 'The root of shame,' Williams writes,

> lies in exposure in a more general sense, in being at a disadvantage: in what I shall call, in a very general phrase, a loss of power. The sense of shame is a reaction of the subject to the consciousness of this loss: in Gabriele Taylor's phrase . . . it is 'the emotion of

self-protection', and in the experience of shame, one's
whole being seems diminished or lessened. In my
experience of shame, the other sees all of me and all
through me . . . the desire [is] to disappear, not to be
there . . .

To put it perversely, shame frees one to disappear; and
frees one to wonder why a loss of power might not be a
consummation devoutly to be wished. Williams implies
that the problem is as much one of exposure as of the
doing of disreputable things; and it is indeed one of the
terrors of shame that it estranges one from others, that
it instigates separation and abandonment (or, to put it
differently, shameful acts are the ultimate love tests).
But shame, as generally conceived, can only be experienced as loss, a devastation exposed, a violation of
privacy. Even if the shameful act was begun in a state
of triumphalism – an exhilarating transgression –
shame is the defeat occasioned by such a triumph
(in shame something has always collapsed). In shame,
though, all one's attention, as Williams suggests, is
given over to a dreaded exposure – 'the other sees all of
me and all through me' – and then hopefully towards
restitution. The all of me that is seen – the all of me to
which attention is being paid – is entirely contemptible;
and so I am utterly defined by what is shameful about
me. It is the 'emotion of self-protection' that presumably
prompts the recovery of a valued self (a reminder of the

self that is simultaneously an attack on the self). But so absorbing is the experience of shame that it prompts the wish to disappear, to not be there. First there is the unbearable attention to the self, and then the murder, the vanishing act of the mortified self. Attention to the so-called external world is long vanished, reduced to a chorus of the appalled. In shame there is a loss of interest in the external world, and a loss of interest in what is or may be good about the self. Shame is the ethnic cleansing of the self. It is a state of horrified and horrifying conviction. It is the making of a dreaded spectacle of oneself.

Shame, that is to say, unlike guilt, is peculiarly difficult to get round; you can't be talked out of feeling mortified (there is more to self-betrayal than simply breaking rules). Indeed, shame seems to be one of the most real things we ever feel ('real' meaning difficult to redescribe, or to transform into something else). In one of the standard psychoanalytic accounts of the difference between shame and guilt – Sandler, Holder and Meers's paper 'The Ego Ideal and the Ideal Self' – guilt is taken to be reparable in a way that shame might not be. 'Shame,' write the authors, 'may be related to "I cannot see myself as I want to see myself or as I want others to see me." Guilt, on the other hand, would be associated with "I do not really want to be what I feel I ought to be." ' Guilt is the consequence of a protest, shame of a failure; where there is shame, there is unassailable consent to the ethical standards that have been violated, but where there is guilt there has been doubt about the rules that have been broken. When there is guilt, there is ambivalence about the rules; when there is shame, there is, perhaps, unconscious ambivalence, an ambivalence that has had to be repressed. We will come back to the idea of shame being, as it were, a more radical protest (against the rules, against presumptions about the preferred self); as well, of course, as the affirmation of moral goodness whose essential value we prefer.

But when we are ashamed of ourselves, we know what we have done and we know that it is terrible; it was already known to be terrible. And in this sense, of course, shame, it would seem, is above all reassuring. Shameful acts do not bring us any moral news; they are not in any obvious sense revisionary (if you are incontinent in public, people may be sympathetic but they won't be impressed). So shame reassures us that we are creatures with integrity and control – creatures with absolute or 'core values' – who suffer most when we let ourselves down (or are seen to be letting ourselves down); our suffering here being a sign of our fundamental commitment to being seen by others and ourselves as we want to be seen, of what we feel when we betray our essential (and known) morality. But then, if it is a question of what is seen, of what is visible – and all the definitions of shame involve something being

41

seen – shame is what we feel about whatever it is about ourselves that we can, and want to, successfully conceal. Shame, in other words, is conservative in the fullest sense; it is there to conserve our privacy, to conserve ourselves as we would prefer to be seen; to conserve our belief that we can conceal ourselves. Because shame is what is felt when we fail to perform well enough (think of the potential humiliation of the stand-up comic). If shame is always possible, performance is everything. Our capacity for shame is a horrifying reminder of what we believe our preferred self really is – and that there is one – however much of a strain it may be to sustain it.

So in one sense, Miller is making a familiar point: that our lives are organized to manage whatever it is about ourselves that we are ashamed of. That to all intents and purposes we are what we can do with, and about, our shame. And this can make concealment, in its various cultural guises, our primary preoccupation; though what is being concealed can also be an all too troubled relation to our cultural ideals, to our preferred versions of ourselves, those objects of desire that have been procured for us ('before I go to the continent I must go to the incontinent', Byron wrote in a letter). Austen's style, in Miller's view – the most distinctive thing about her – comes out of a 'shameful relation' to marriage, something that nobody in her world could be indifferent to. We are more likely now to have a shameful relation to money, or sex, or food, or fame, or politics; but it is only possible to have a shameful relation to the things that matter most. That is how we know what matters most, or what is supposed to. By being ashamed of ourselves, we reveal what we value about ourselves, and how we avoid our doubts about what we value; as though shame organizes both our taste and our moral intelligence. So essential is the shame we need to conceal that it informs everything. Our style is always a shameful relation to something. Or that, at least, is what Miller might want us to consider.

Clearly, when we are ashamed of ourselves there is something we have failed to be, or to do, that is deemed essential; and this failure to be or to do something or other has been exposed (nakedness is the figure for shame, because there is no other body that we can be). But in our falling short we can wonder whether it is ourselves or our standards that are at fault, or whether more simply it is a travesty (though it is this questioning, as I have said, that guilt makes possible and shame disables). And we can also wonder, by the same token, how we might recruit our ideals in our quest for shame (shame successfully disarms a scepticism about our cultural ideals that we may want to recover). Though we tend to talk of people sometimes enjoying their guilt, we rarely talk of their enjoying feeling ashamed – and we should. Not merely because of the exquisite and excruciating punishment shame provides (along with

43

the relief of compromised attention), but because it is also our most visceral protest against our most demeaning confinements. When shame is not sustaining our values, it is pre-empting our capacity for moral enquiry. And in this sense it is always an uncompleted action; it forecloses the question: what else could happen next, other than the appropriate and apparently deserved mortification?

That we are capable of feeling something we call shame – that we are prone to feeling mortified – might suggest that there is something, or many things, in the feeling of shame that we want, or even need. That it is a way of saying something that has not, as yet, found another form; which means that shame may be the thing we are least able to adequately respond to, in both ourselves and others. We can suffer our shame – and that of others – but we can't read it. We can punish it but we can't easily and interestingly interpret it. We can't imagine what else – what unknown future – it could lead to. It is our relation to shame that is shameful, not shame itself.

If we described shame as an exposé, and not only or solely a confirmation, of our ideals (and of our willingness and ability to live up to them: it is remarkable the lengths to which people will go to risk being shamed, to have the opportunity of being shamed); or as a baffled and therefore baffling request for something other than punishment and correction (rather than a state of conviction); or even as a pre-empting of curiosity (hiding from a puzzling experience in a false certainty), we could at least short-circuit the shame there always is in talking about shame. This is perhaps what Edmund intimates, in talking to Fanny in Austen's *Mansfield Park*, when he refers to 'a mixture of many feelings – a great though short struggle – half a wish of yielding to truths, half a sense of shame'. This shame, he implies, is a refuge from the yielding to truths and from the mixture of many feelings that make shame a gripping oversimplification. It is a short struggle because shame is the abiding temptation. There is something irresistible about it. And this too is worth wondering about: what is it about shame that is at once so alluring and persuasive?

We don't now think of Jane Austen as merely refusing or failing to get married, any more than we think of her novels as simply promoting or endorsing marriage (or the marriage market). If Miller is right, Austen's was a shameful relation to marriage that, like many shameful relations, was potentially a new beginning. What may have been shame in the first instance opened up into the strangely expansive world of her novels. To transform shame into style is not to settle for it.

II

The trouble with a secret life is that it is very frequently a secret from the person who lives it and not at all a secret for the people he encounters.

James Baldwin, *Another Country*

In James Baldwin's *Giovanni's Room* — another novel about the 'conjugal imperative', about the kind of attention the marriage question requires — a man discovers, or rediscovers, his homosexual desire in the process of gradually separating from a 'girlfriend' whom he intends to marry. It is a book in which, perhaps unsurprisingly, shame is used, as it were, as a sign of the real (perhaps a legacy of Baldwin's religious upbringing). Baldwin's characters feel shame when they are not being true to what is most valuable, and alive, in their lives. When shame is referred to, the book becomes what we might think of as moralistic in the nicest possible way; the characters are revealed as having some kind of moral compass, some assured sense of the true and the good. The novel is otherwise acute about confusion, and about how terrified people are of how they feel about each other; about just how disturbing and intriguing it is that we confound and betray ourselves and each other as we do. But when shame turns up, we know where we are. As though moral dilemmas can be resolved by moral truths. As though when there is shame we know where we are, and perhaps who we are.

Baldwin's David has a louche, wealthy older gay friend called Jacques, who is notorious in their circle in Paris for his disreputable encounters with younger, poorer boys. 'Tell me,' David says to Jacques one evening,

'is there really no other way for you but this? To kneel down forever before an army of boys for just five dirty minutes in the dark?'

'Think,' said Jacques, 'of the men who have kneeled before you while you thought of something else and pretended that nothing was happening down there in the dark between your legs . . . You think,' he persisted, 'that my life is shameful because my encounters are. And they are. But you should ask yourself why they are.'

'Why are they – shameful?' I asked him.

'Because there is no affection in them, and no joy. It's like putting an electric plug in a dead socket. Touch, but no contact. All touch, but no contact and no light.'

I asked him: 'Why?'

'That you must ask yourself,' he told me, 'and perhaps one day this morning will not be ashes in your mouth.'

Jacques begins by suggesting that what seems obvious is not self-evident; that it involves questions even if, or particularly if, it doesn't seem to invite them, or need them. He prompts David to ask what is shameful about shameful acts; and by implication to think: if these are shameful acts, what qualifies an encounter as being without shame; or rather, more ambiguously, as shameless? Of course his life is shameful, he says, because his encounters are – as though a life is just a series of encounters – 'But you should ask yourself why they are.' In other words, David should ask himself rather than Jacques what it is that makes these encounters shameful. The army of boys suggests conscription; and Jacques knows what is shameful about his life, though not necessarily, any more than anybody does, quite what makes these things shameful. And yet like all people who are ashamed of themselves, he knows exactly what really matters; his shame discloses his moral clarity (a moral clarity he may have intermittently dissociated himself from). There is no affection, no joy; there is touch but without contact or light; without enlightenment or illumination. And most readers will agree that no one promotes or celebrates the absence of affection, or joy, or contact, or light. But in this there is, it should be noted, an elaborate story about sexuality, and about what is of value between people. And there is no debate about this; an already existing consensus has been recovered. In what Caryl Phillips has called Baldwin's 'audacious second novel' – 'a novel by a young black writer containing no black characters, and dealing with the taboo subject of homosexuality' – it is shame that somehow organizes the narrative, in part by alluding to its tensions and omissions. Audacity always flirts with shame.

It is, pointedly, not the shame cultivated and contrived by racism and homophobia that Baldwin refers to, but the shame of impoverished human relationships. There is shame in *Giovanni's Room* when there is deprivation – deprivation felt as demeaning; the deprivation that renders people abject and self-hating rather than personally and politically enraged and engaged. Shame is not experienced by Baldwin's characters as injustice, but as a confounding lack of relatedness; there is no affection, no joy in the encounters: 'It's like putting an electric plug in a dead socket.' There is no connection (the allusion may be to Whitman's 'I Sing the Body Electric'). What is, I think, especially poignant and compelling in the novel is the narrator's sense that whatever else shame is – and it may be many things, 'a mixture of many feelings', in Austen's words – it refers to lack of fellow feeling; what Baldwin calls, winningly, 'affection'. It is not sexuality that is resisted, but the

49

affection in it. It may be mortifying to feel so estranged when there is the possibility of intimacy ('there is no sex without love or its refusal', Baldwin's gay contemporary the American writer Paul Goodman said). As David is thrown into a state of sexual frenzy walking round Paris, he notices a sailor:

> But hurrying, and not daring now to look at anyone, male or female, who passed me on the broad sidewalks, I knew that what the sailor had seen in my unguarded eyes was envy and desire. I had seen it often in Jacques' eyes and my reaction and the sailor's had been the same. But if I were still able to feel affection and if he had seen it in my eyes, it would not have helped, for affection, for the boys I was doomed to look at, was vastly more frightening than lust.

When you become unguarded, the guard of shame turns up; it quite literally guides and organizes your attention, what you can let yourself see. In the uncanny game of seeing and being seen, the envy and desire – the 'lust' – is not the terror; it is the affection that is 'vastly more frightening'. Shame, paradoxically, is a punishment for, or a refusal of, fellow feeling, for the acknowledgement of vulnerability, which is always a vulnerability shared. Shame then is estranging, and also an estrangement technique. When we are ashamed of ourselves we are effectively saying, 'this is what I really am,' and, at the same time, 'this can't, this mustn't be me'; which like all not-me experiences is an affinity disowned. The person who is ashamed of himself confirms and affirms a community in the act of estranging himself from it: this is what he draws our attention to. When Baldwin was asked in his *Paris Review* interview of 1984, 'Did what you wanted to write come easily to you from the start?', he replied, 'I had to be released from a terrible shyness — an illusion that I could hide anything from anybody.' So we might think of shame as a terrible shyness. Baldwin suggests in *Giovanni's Room* that David's shame of being gay is his shame of feeling affection for men, for which the lust is both an elaboration and a cover story. And then, as always, there is the further shame, of lying to oneself and others, in this case about one's affections. In David's final separation from the woman he is to marry — the aptly named Hella — she says to him, 'But I knew . . . I knew. This is what makes me so ashamed. I knew it every time you looked at me. I knew it every time we went to bed. If only you had told me the truth then. Don't you see how unjust it was to wait for me to find it out?' If Hella is ashamed of her knowing self-deception, she must want to be — or believe herself to be — someone who tells herself the truth about her feelings and perceptions; and someone who knows — who has the capacity to recognize — these truths. Shame, then, is also a sign of self-deception, and the fear of it an abiding temptation: a baffling or a violation of a well-known morality. If you want to know what you want to

be, pay attention to what makes you feel ashamed of yourself. It is a revelation, so to speak, of your ego-ideal (who
you, in the fullest sense, wish to be: how you would like
to look). But it is also, by the same token, a revelation of
an assumed self-knowledge (someone who had no idea
who they were or wanted to be could not feel shame).
Either Hella knows what she feels (and perceives), or she
wants to be the kind of person who does. And if she
knows, she wants to be the kind of person who can act on
her knowledge, who doesn't wishfully set it aside.

'Behaviour that is in conflict with the super-ego,' the
psychoanalyst Charles Rycroft writes, 'evokes guilt, while
that which conflicts with the ego-ideal evokes shame.'
Hella is ashamed because in her ideal picture of herself she
is someone who trusts what she feels. In the Freudian
story, broadly speaking, the super-ego represents the
internalized rules of the parents, the ego-ideal the internalized, preferred version of oneself: 'a model to which the
subject attempts to conform', according to Laplanche and
Pontalis's *The Language of Psychoanalysis*; making us
wonder what it might be to attempt to conform to a model,
and about this being a picture of development. 'The ego-
ideal,' writes the Italian analyst Antonino Ferro in *Seeds
of Illness, Seeds of Recovery*, '. . . performs functions of protection and stimulation . . . [but] may also be pathologically
tyrannical, peremptorily insisting on the achievement
of high and unattainable objectives . . . [It] not only gives
rise to frustration but also exposes the subject to

self-devaluation . . . [it functions characteristically with] mildness or the imposition of malignant demands'. The ego-ideal, that is to say, creates an essential perplexity. Is it safeguarding and sustaining what is best about us – what we value above all about ourselves, about life – or is it humiliating us with unattainable ideals? And are we able to think about and talk about the values it promotes, or have we been, as it were, brainwashed or enchanted by our ideals? Are we suffering from our obedience – our indebtedness – or for our values? If, for example, we started from the principle that no one owes anyone anything, we could begin to think about what we want to give each other. Are we setting ourselves up to fail, or bringing out the best in ourselves?

It is clear from this cursory albeit psychoanalytic account that to question one's preferred version of oneself – the model of oneself one attempts to conform to – is a quite different, though linked, project to breaking the rules and the rule of the parents. And this would be a good description of two kinds of heroes and heroines that novels are good at thinking about. Hella and David in *Giovanni's Room* are not worried about breaking rules; they are disturbed about not living up to their most cherished ideals for themselves. We should note in passing that the ego-ideal described as 'a model to which the subject attempts to conform' involves, as I say, the idea both of a model and of conformity; and then a question about what kind of attention such models may require of us (devotional, sceptical, entranced, ironic?). There is a picture to be realized, projects to be accomplished, however unconscious, in the shaping of the self. So shame, as I have been suggesting, raises in an acute form questions about the power and strangeness of identification – of attempting to conform to a model – in the shaping of a self. It represents the self, by definition, as always aspiring, always in a state of emulation, never quite itself because always as yet unattained. And yet this same self can claim to know what it wants to be. Shame measures the distance between who we experience ourselves as being, and who we would like to be – the distance between our ego and our ego-ideal that is the source of our suffering.

What *Giovanni's Room* helps us to see is that though we may be very conscious of our preferred version of ourselves – indeed, we may, to all intents and purposes, be thinking of nothing else – it may also be true that only shame more fully discloses or reveals the model to which we are attempting to conform (and also the enigma of this wanting to conform, as though the self was merely its repertoire of more or less successful imitations). There is a difference between the models we are conscious of, and those we are not; a difference between our conscious and our unconscious affinities. In other words, we may unconsciously seek out shaming experiences to realize in our falling short just what it was we had been attempting or wanting to attain.

There is clearly a tyranny – a perhaps more insidious or subtle tyranny – of the unconscious model of ourselves that we are attempting to conform to. I may be conscious of trying to conform to a heterosexual model of myself that is in conflict with my homosexual desire. I may be unconscious of wanting to be able to be affectionate with men, or of wanting to be truthful with myself and others about what I experience. When I am thoroughly ashamed of myself there is the possibility of such clarities. But shame, as I say, always runs the risk of being an uncompleted action; one that supposedly should lead back to a known and abandoned ideal self rather than forward to a newly reconceived self. To have a shameful relation to anything – marriage, homosexuality, money – is to be working out, in however baffled and baffling a way – the unconscious models one may be trying to conform to. And this brings with it, of course – at least to the enlightenment-minded – the possibility that in becoming aware of something, one might be able to change it. Or more simply, one might be able to have different kinds of conversation about it. Shame may be more of a confounding of sociability than a despair about it.

At one point Giovanni, the 'hero' of the novel, turns to David and looks at him: 'His eyes were red and wet, but he wore a strange smile, it was composed of cruelty and shame and delight.' Baldwin is reminding us here that shame is compatible with what Austen called 'a mixture of many feelings', and that it may be itself a composition of feelings. A lot of work goes into the attempt to conform to a preferred model of oneself; a lot of the work being the work of representation itself, the finding and making of alluring figures to conform to.

III

What both biologically and psychologically defines life
is that it is transmitted.

> J. B. Pontalis, 'The Birth and
> Recognition of the Self'

In the psychoanalytic story, it is (notoriously) sexuality
that forms, informs and deforms our attempt to conform
to a model; and that most intimately involves us in questions about how we want to be seen, by ourselves and
others (sexuality, it should be noted, is not Freud's Big
Idea but an array of prolific small ideas). The models of
course are culturally transmitted, the drives biologically
transmitted. And so there is always an easy link to be
made between sexuality and shame. And yet when we
turn to Freud, the author of the essential description of
sexuality in psychoanalysis, we find a notable absence of
interesting accounts of shame. It is as though shame is
something that Freud, like the rest of us, is unwilling or
unable to elaborate theoretically, while insisting on its

significance in sexual development. In his *Three Essays on the Theory of Sexuality*, for example, there are thirteen references to shame, virtually all of which reiterate, often verbatim, the same point, repetition, of course, being something that Freud is keen to wake us up to, and to wake himself up to (in shorthand we can say that for Freud, where there is repetition there is unfinished or unfinishable business; shame is always repetition without improvisation; where there is shame there is always something traumatic, that is something – like a core belief – apparently untransformable). Not only does Freud keep saying the same thing about shame in the *Three Essays*; what he does say is often – and has become – the stuff of psychoanalytic cliché. Shame is a defence against sex; it exposes furtive and mortifying desire. And to add insult to injury, James Strachey's editor's note to the 1949 edition gives the game away before we have even started reading Freud. In Freud's early letters to his colleague Wilhelm Fliess, Strachey remarks, we can see the first 'indications of a more psychological approach – a discussion appears of the repressive forces, disgust, shame and morality'. Shame, then, as one of Freud's repressive forces; interestingly categorized with disgust and morality though clearly these can also be three very different things, and not merely or solely 'repressive forces' (shame is morality at its most fundamentalist, and aesthetics at its crudest). By describing it as one of the repressive forces, Freud is giving it a kind of agency; as though our shame does something to or for us; it acts on our behalf. We have to note what Freud associates shame with to get a sense of what he might be getting at.

In the section on 'Perversions' in the *Three Essays*, Freud refers to 'the sexual instinct [going] to astonishing lengths in successfully overriding the resistances of shame, disgust, horror or pain' (he cites as examples 'cases of licking excrement or of intercourse with dead bodies'); as though the sexual instinct has to be imaginative in a way our defences can't be. 'Certain mental forces,' he continues, 'act as resistances, of which shame and disgust are the most prominent'; 'most prominent' meaning the most noticeable and/or the most effective. Hysterics, he writes further on, because they are excessively (in Freud's view) sexually repressed, show 'an intensification of resistance against the sexual instinct (which we have already met with in the form of shame, disgust and morality)'. Under the heading 'Sexual Inhibitions', Freud writes of what he calls latency as the period of development in which 'are built up the mental forces which are later to impede the course of the sexual instinct, and, like dams, restrict its flow – disgust, feelings of shame, and the claims of aesthetic and moral ideals', 'built up' in the sense that the individual is actively building his protection from his instinctual life; it is a made thing. He refers later to 'shame, disgust and morality' as 'mental dams against sexual excesses'; but these constructions are, as it were, developmental achievements. 'Small children,' he writes, are essentially without shame, and at some periods of their earliest years show an unmistakable satisfaction in exposing their bodies, with especial emphasis on the sexual parts. The counterpart of this supposedly perverse inclination, curiosity to see other people's genitals, probably does not become manifest until sometime later in childhood, when the obstacle set up by the sense of shame has already reached a certain degree of development.

'The masculine and feminine dispositions are already easily recognizable in childhood,' he continues. 'The development of the inhibitions of sexuality (shame, disgust, pity, etc.) take place in little girls earlier and in the face of less resistance than in boys.' And then by way of summary and conclusion, he writes: 'Among the forces restricting the direction taken by the sexual instinct we laid emphasis upon shame, disgust, pity and the structures of morality and authority erected by society.' I cite all these somewhat tediously repetitive examples as a way of asking a question, Freud having alerted us to the provocation of insistent repetition. What is Freud saying, and why does he keep saying it? What are his formulations a shameful relation to?

As I say, shame for Freud is in a category with disgust, pity, aesthetic and moral ideals, the structures of morality and authority erected by society. It is, like them, a form of attention, at once coerced and coercive;

it, as it were, makes us see things in a particular way. We could say it is the kind of attention we use when we want to mortify ourselves; each form of attention being a way of getting to a certain kind of experience. Disgust and pity, aesthetic and moral ideals are all described by Freud as forces, 'mental forces', that dam and inhibit and restrict and impede and obstruct the sexual instinct, and what he calls its flow and direction; they are 'resistances' to sexuality. Everything in his argument is reactive to sexuality; everything is a way of coping with desire (a failing or a 'refusing', in D. A. Miller's language; or a kind of sham). The excessively sexually repressed, the hysterics, require more inhibitions like shame to keep them safe from desire. Children, unlike the hysterics, are 'essentially without shame', which means they enjoy exposing their bodies, and particularly their genitals; and are easily and pleasurably curious about other people's genitals (so shame is also an inhibitor of that essential ingredient of sexuality, curiosity; it is a form of attention that forecloses attention; the saboteur of curiosity, both about itself and about what it has made us ashamed of). And one of the supposedly essential differences between boys and girls, observable in childhood according to Freud, is that the sexuality of girls is inhibited earlier and with less resistance than that of boys. Shame, like all the other resistances to sexuality, is deemed to be a developmental achievement, a prerequisite of acculturation.

For Freud, then, it is one of those essential 'forces' without which sexuality would flow more freely, and so more disruptively, and thereby endanger the individual and his culture. Sex requires shame just as water, from some people's point of view, needs to be dammed (the pun is only in English). Without the achievement of shame, it is implied, we might go on exposing our bodies and so our genitals and being wholeheartedly curious about other people's (or, we might say, about other people). Or at least we would certainly be tempted to. Where would it all end?

To put it slightly differently, it may be one of the functions of shame to pre-empt the consequences – or specifically to pre-empt *imagining* the consequences – of forbidden acts. Shame dams up our thinking, our imagining, our conversation. And this, of course, depending on one's point of view, may be no bad thing. In a world without shame, there might, for example, be even more bullying, even more intimidation, than there is now. To put it differently, if you think of yourself as wanting more sex, what exactly is it that you are wanting more of? In these formulations Freud is implicitly inviting us to imagine and describe what the less inhibited sexual – and aggressive – life would look like. And indeed the way he puts it suggests that shame is at least one key or clue to our desire; it is what we are likely to feel – the resistance we will put up, the dam we will erect, the moral or aesthetic ideal we will promote – when there is a compelling but forbidden desire in the offing, or being kept at bay. In this story we can only apprehend our desire by decoding our resistances to it. Sexuality must be daemonic if it requires such elaborate cultural work to contain it. The greater the shame, the greater the urgency to disrupt desire. And of course women must be, given Freud's account of them here, as it were, more sexual and so have more to be ashamed of than men (the men's shame being delegated to the women); or, of course, more civilized, more developed (the men's ambitions for themselves delegated to the women). If nothing else, the link between shame and misogyny should make us think; the way men attribute to women the parts of themselves that disturb them, and then punish the women accordingly. What men want and need to believe about women is itself a source of shame.

I think it is worth noting that by categorizing shame so certainly with moral and aesthetic ideals and structures of authority, Freud is intimating that it might have a complexity – the complexity of structures of authority, or of the aesthetic, say – that the experience of shame may dissuade us from considering. And if shame is akin to moral and aesthetic ideals and structures of authority, it is something people make, and might therefore be able to unmake or remake. Whereas if it is akin to disgust or pity, we may all too easily think of it as natural, as part of us rather than made by us (we change our attention by changing our beliefs). It is precisely these equivocations, these ambiguities about shame, that Freud allows for. But perhaps we should note by way of conclusion, or by inference, that shame itself is the consequence of a form of attention; it is evidence, so to speak, of the kind of attention we are paying to our sexual desire. Where there is shame, it is as though we have seen something very clearly, or assume we have; in Freud's account it is as though because we know what sex is, we can explain shame, not to mention authority, morality, aesthetics and pity. Through shame we pay attention, we have to pay attention, to our most unacceptable desires. And we draw attention to them – 'the other sees all of me and all through me' – in our mortification. But we know what it is; we know what we are up against, and that we are up against something (when I am ashamed I know what I am seeing, I have no doubt. Shame is a cure for scepticism). The shamed person reminds people of what he and they are ashamed of; he reminds them of what they already know. And in that sense shame recycles – perhaps for reconsideration – unacceptable desire. To be ashamed of oneself, then, is the ultimate love test, but of oneself, as well as of others.

In his great book *The Greeks and the Irrational*, E. R. Dodds writes of 'the tension between individual impulse and the pressure of social conformity characteristic of shame culture'. Picking up on the anthropologist

Ruth Benedict's distinction between guilt cultures and shame cultures, Dodds elaborates the ethos of the shame culture. 'In such a society,' he writes, 'anything which exposes a man to the contempt or ridicule of his fellows, which causes him to "lose face", is felt as unbearable.' Shame, Freud writes, is the losing face caused by sexuality; it is sexuality – our desire, and our desire for others – that leaves us all too prone to the contempt and ridicule of ourselves and others (as though shame dictates where we look and how we look). But perhaps, by the same token, Freud's account of sexuality – paradoxically – reveals his shameful relation to curiosity, or even to sexuality itself, which for him is the source and aim of curiosity. After all, he has warned us in no uncertain terms about our life task of protecting ourselves from our desire ('Man's project,' Lacan famously pronounced, reinterpreting Freud, 'is to escape from his desire'). So Freud himself, in his text, can hardly be exempt from this project. In order to defend yourself against sex, you have to know – or believe that you know – what it is. Freud, I think, has a shameful relation to his omniscience about sexuality: he fears, like everyone else, his naivety about sex. This is what he wants to conceal and expose in his great work of psychoanalysis. He fears the contempt and ridicule of his fellows; he fears the exposure of his ignorance and his curiosity around and about sexuality. And he was right: it was to be sexuality that created the controversy about psychoanalysis. The pleasure and the terror of sexuality is the continual risk, and possibility – and perhaps pleasure – of losing face. Of being paid, and paying, the wrong kind of attention; of wanting the wrong kind of attention. Shame exposes the tyranny of the face we must not lose.

Vacancies of Attention

I

To be observed, to be attended to, to be taken notice of with sympathy, complacency and approbation.
Adam Smith, *The Theory of Moral Sentiments*

In 1920, Freud added a passage from *Tristram Shandy* to the chapter 'Symptomatic and Chance Actions' in *The Psychopathology of Everyday Life*. 'In the field of symptomatic acts too,' he wrote in one of many such acknowledgements in his work, 'psycho-analytic observation must concede priority to imaginative writers. It can only repeat what they have said long ago.' If psychoanalysis can only repeat what imaginative writers have said long ago, it is as though lessons haven't been learned, that we have been insufficiently attentive readers, or that we are unduly resistant to the provocations of imaginative writers. It is psychoanalytic observation that must concede priority presumably to the observation of what Freud calls 'imaginative writers'. It is the quality of attention and the need for repetition – and the links between attention and repetition – that exercises him here, and, indeed, throughout *The Psychopathology of Everyday Life* (repetition is always repetition of attention). The subtitle of the book – *On Forgetfulness, Slips of the Tongue, Inadvertent Actions, Superstitions and Mistakes* – is a catalogue of inattention and its discontents.

After his preamble, Freud quotes something that he says was 'drawn to his attention' from Sterne's novel. 'I am not at all surprised,' Tristram's father says,

> that Gregory of Nazianzum, upon observing the hasty and untoward gestures of Julian, should foretell he would one day become an apostate; – or that St Ambrose should turn his Amanuensis out of doors, because of an indecent motion of his head, which went backwards and forwards like a flail; – or that Democritus should conceive Protagoras to be a scholar, from seeing him bind up a faggot, and thrusting, as he did it, the small twigs inwards. – There are a thousand unnoticed openings, continued my father, which let a penetrating eye at once into a man's soul; and I maintain it, added he, that a man of sense does not lay down his hat in coming into a room, – or take it up in going out of it, but something escapes, which discovers him.

What has been drawn to Freud's attention, that he is now drawing to our attention, is Sterne's account of paying a certain kind of attention; attention to the 'thousand unnoticed openings . . . which let a penetrating eye at once into a man's soul'. And, by the same token, to what may be missed either by inattention or by not paying this kind of attention; to the consequences of the not-noticings of everyday life. But without the penetrating eye, there are no openings; it is the penetrating eye that is the precondition, in Sterne's overtly sexual image, of such openings. And Freud's book is an account of the new penetrating eye – and ear – of the psychoanalyst. In retrospect we can say that Freud realized that his patients were in search of a new kind of attention, a new kind of attention for their inattentions. If you believed, as he was beginning to do, in the value and efficacy of psychoanalytic treatment, you could say that their symptoms were provocations that had misfired; that they were (unconsciously) addressed to a different kind of doctor. And this new doctor, as we shall see, was defined by the quality of his attention, and of his inattention.

Freud is describing in these early formative books – the essays on sexuality, the joke book, *Interpreting Dreams* and *The Psychopathology of Everyday Life* – what he takes to be a new kind of attention, psychoanalytic attention; an interpretative attention that is in the service of telling and useful descriptions of unconscious motivation. But what is perhaps also of interest in Freud's use of *Tristram Shandy*, and which Freud doesn't comment on – he quotes Sterne without commentary, as though his words speak for themselves – is that Sterne's examples are about predicting the future, whereas Freud's examples in *The Psychopathology of Everyday Life* are about predicting the past; they are conjectural descriptions of desires from the past that are active in the person's present life. Slips are uncompleted provocations. In actuality, Freud's new science makes very few, if any, predictive claims about a person's future. Psychoanalytic treatment is about the unpredictable consequences of certain evolving acknowledgements. All psychoanalysis can reveal is the nature, which is in part the history, of the patient's desire. In and of itself, the psychoanalytic attention given to this cannot, by definition, disclose anything about the patient's future.

Tristram's father suggests, not unlike Freud, that there is virtually nothing a person does that is not remarkably revealing – 'a man of sense does not lay down his hat in coming into a room – or take it up in going out of it, but something escapes which discovers him'. A 'man of sense' because as (eighteenth-century) sensible creatures, what people do makes sense; and however guarded we may be, something escapes that 'discovers' us, that reveals us; as though, whether we want to or not, we make something of ourselves known; we give ourselves away. This giving ourselves away is a making sense; sense can be made of it, and it can make sense. Our inattention invites attention. So we are more communal, more communicative, more potentially in contact with others than we are always aware of, or want to know about (we are always more sociable, more communicative, than we realize). What Freud calls 'symptomatic acts' – many of which are, of course, acts of apparent inattention, like slips and bungles – may be a kind of love test to the world; a testing of attention and engagement (as though symptomatic acts are like jokes, something people either get or they don't; as though acts of inattention are courting attention: are themselves provocations). So the question is raised by Freud: are so-called symptomatic acts an (unconscious) attempt to provoke a certain kind of attention, or does a certain kind of attention make them symptomatic acts? Are they provocations by intention, as it were, however unconscious: or does a certain kind of attention turn them into provocations? When is provocation in the eye of the beholder, and when does it expose the eye of the beholder? Freud is interested in the mistakes people make, and in what can be made of what people make. Clearly a slip of the tongue made while buying a newspaper is different from a slip made to one's analyst. And the phrase a 'man of sense', of course, meant something different in the eighteenth century than it does, or could, now.

One way of thinking about this is to wonder what happens to all the symptomatic acts – the huge majority – that are unattended to. What happens, in Sterne's language, if something escapes that discovers a person and nobody notices, which happens all the time? The pragmatic answer is that they go on doing it, or find new ways of doing it, in the hope that someone will notice it. The dismaying answer is that a joke won't amuse you unless it does. In Freud's story it is worth noticing – and we will come back to this – that the privileging of recognition doesn't always tell us very much, or enough, about where, it is assumed, recognition might lead; this recognition that depends on a certain kind of attention. It is easy to believe that underlying our instincts – our putative needs and wants – and indeed the precondition for them, is a need for recognition. Needs without anyone able to acknowledge them are torments; provocations that misfire. But what Freud adds – and not only Freud – is that we are seeking recognition of our inattentions; and that inattention is itself a way of seeking recognition. Inattention – as every child and parent knows – is its own kind of provocation. If Freud was a pragmatist, we could say that he was interested in the uses of inattention.

'Western modernity since the nineteenth century,' writes the critic Jonathan Crary in *Suspensions of Perception*, 'has demanded that individuals define and shape themselves in terms of a capacity for "paying attention", that is, for a disengagement from a broader field of attraction, whether visual or auditory, for the sake of isolating or focusing on a reduced number of stimuli.' Attention is ineluctably selective: it is made

74

possible by inattention; as if, by the same token, as Crary intimates, individuals define and shape themselves by what they fail or refuse to pay attention to. And indeed, it is this question of the reasons for inattention – the reasons being as much of a provocation as the inattentions – that Freud addresses; the question of whether, and to what extent, these inattentions are refusals, or failures, or incapacities (or, indeed, provocations), and what these inattentions might be in the service of. If, as Crary writes, 'the articulation of a subject in terms of attentive capacities simultaneously disclosed a subject incapable of conforming to such disciplinary imperatives', we are left wondering why impossible attentive capacities might be demanded, and what attention has to do with conformity. What Crary calls 'attentive capacities' conforming to disciplinary imperatives always involves the imposition of an essentialism. There is something to which, because of who we supposedly are, we should be attending.

Certainly in Freud's increasingly essentialist view, civilization has its discontents because civilization makes us pay attention to the wrong things; and when we pay attention to the right things – our instinctual life – we may increase our pleasure, though by doing so we also increase our suffering. But before the later and abiding disillusionments of *Civilization and Its Discontents* – and while Freud was still in the process of becoming the committed essentialist he turned out to be – the younger Freud's question was: what are the preconditions for inattention? And then: to what can inattention lead? One thing inattention could lead to was psychoanalysis. It was these provocations of inattention – what kind of attention the inattentions of his patients called up in him – that Freud would elaborate on. Psychoanalysis, as we shall see, was to be a language that cultivated an interest in, and a commitment to, inattention, and the kind of attention and inattention we can bring to it. But I want first to return to *Tristram Shandy* – by way of Samuel Johnson's contemporary fable *The History of Rasselas* – for earlier and instructive ways of describing the uses of inattention; ways that also surface, I think, wittingly or unwittingly, in the work of Marion Milner, a member of what became known as the Middle, or Independent, Group in British psychoanalysis.

II

The world is like nothing we've ever seen.

Emily Hasler, 'Wet Season'

Samuel Johnson's *The History of Rasselas, Prince of* Abissinia begins with a description of a palace in a 'wide and fruitful' valley where the emperor's children live until the 'order of succession' calls them to the throne. It is described as an idyll in which 'the blessings of nature were collected, and its evils extracted and excluded'. A kind of eighteenth-century orientalist gated community, it is a place of peace, serenity and beauty, without challenge or threat. Once a year the emperor would visit his children; and during the eight days of his visit,

every one that resided in the valley was required to propose whatever might contribute to make seclusion pleasant, to fill up the vacancies of attention, and lessen the tediousness of time. Every desire was immediately

77

granted. All the artificers of pleasure were called to gladden the festivity.'

It is clear that all is not well in the valley. Something must be lacking if there is such a need to lessen the tediousness of time. What is the lesson in this lessening that seems so urgent? The hope is that the artificers of pleasure will do the trick, and yet each year the emperor returns and the malady is once again addressed. It never goes away. It is, of course, the danger of living apparently satisfied in an over-organized environment that Johnson is so pointedly alerting us to at the very beginning of the story; the terrors of some versions of social engineering, in which we are entombed within our supposed preferences and ideals. What is absent in what is called notably the Happy Valley is the tension of desiring; the freedom to think about what is missing, and what might be missed (the inventiveness born of frustration). Or as Johnson suggests, in what we might think of, wrongly, as a more contemporary vocabulary, enjoyment can be used to pre-empt desire. 'I have already enjoyed too much; give me something to desire,' Prince Rasselas declares in a chapter entitled 'The Wants of Him that Wants Nothing'. It is the ways in which satisfaction can sabotage desire that Johnson wants us to think about in his moral tale; and he connects this – as we do in the modern language that is psychoanalysis – with attention and its vacancies. So much depends on where our attention is. If acculturation is, among other things, the organizing of attention – or the organizing of desire as the organizing of attention – then there is a tension, as Johnson implies, between what we are supposed to attend to, and what we find ourselves wanting to attend to. We have an emperor, sages, and artificers of pleasure to help us with our vacancies of attention (Johnson – the only great writer who wrote a dictionary, and who thus provokes a unique attention to his words – defines 'vacant' as 'empty, void', but also as 'free, unencumbered'; as 'thoughtless, empty of thought', but also as 'being at leisure'). As though our vacancies of attention were not one of the forms our attention takes.

Rasselas has to leave the Happy Valley. 'He resolved to obtain some knowledge of the ways of men,' the one thing he couldn't find or find out about there. It was, Johnson writes, the prince's 'curiosity', 'a source of inexhaustible inquiry', that got him out; a desire for a certain kind of knowledge – a knowledge of other people, strangers to the Happy Valley – and a question about why such knowledge was wanted. Freedom, or rather change, is described as a shift in attention; and vacancies of attention are the precondition for change. Thus we need to be alert, Johnson intimates, to the ways in which systems or regimes or vocabularies try to pre-empt vacancies of attention; and to what kind of vacancies of attention they tend to incite.

'Vacancies of attention' is a phrase worth attending to, not least because it suggests that at leisure, unencumbered, we are in a different kind of elsewhere. In such vacancies significant realizations may occur, gaps in knowledge may be revelatory or inspiring or confounding, other desires may float into view. It makes us wonder what we are doing when we are filling up the vacancies of our attention, and what attention might be, or be like – how we picture it – if it requires filling, or can empty (as though we equated attention with time). And if filling, as Johnson defines it in his dictionary, is 'to make full . . . to engage to employ', what then is empty, or unengaged, or unemployed attention? What are we doing, if anything, in the vacancies of our attention? These are, of course, the same questions that will later inspire psychoanalytic enquiry. But it is almost as though Johnson is wondering here what or where attention is when it isn't there; or more pragmatically, what happens to attention – to what he calls 'desire' – when there is nothing to organize it, nothing sufficient for it to focus on.

Once we are invited to imagine the absence of attention – attention as something that can be absent, or that can absent itself – attention itself becomes more perplexing. The vacancies of attention suggest that there is, as it were, more than one of them; and indeed that our attention may be also beyond our control; or that control might be the wrong word, as it often is, to use about our attention. The faux optimism at the beginning of *The History of Rasselas* is that with the arrival of the emperor, these vacancies of attention can be filled. As it turns out, it is only Rasselas's 'curiosity' that can do the trick. In his dictionary Johnson defines an act of curiosity as a 'nice experiment'; to be curious, he writes, is to be 'attentive to . . . exact, nice, subtle'. There are vacancies of attention, and there is the nice experiment of curiosity to which they can lead.

'To attend', Johnson writes, is 'to fix the mind upon', with both meanings of fix in play; the mind is repaired and organized, calmed or stilled by concentrating on something external. And then there is the other meaning of 'attend', which is somehow complementary: 'to wait on, to accompany as an inferior, or a servant'. The mind, which always runs the risk for Johnson of becoming unfixed, unmoored, depends like a servant on his master the external world that it serves; a world created by God, and to which it must attend. The external world keeps us sane; the internal world for Johnson is a potential fall into madness, into delirium. His terror is of the tyrannies of the secluded mind, so attention is something he must always keep his eye on (in his dictionary he defines an 'attender' as a 'companion' or 'associate').

The critic Christian Thorne instructively links Johnson's filling-up of the vacancies of attention to lessen the tediousness of time with Sterne's contemporary notion of the hobby horse in *Tristram Shandy*. It is a useful link because Sterne is as preoccupied with states of attention as his contemporary Johnson is; preoccupied by the ways in which we can and cannot in some new sense choose both the object and the quality of our attention, something that is both a modern and a provoking preoccupation. 'Digression,' as the critic Matthew Bevis has written, 'being Tristram's signature tune and the life of the book': 'another name for inattention and distraction'. Sterne's Uncle Toby, and his absurd and absorbing obsession with a war he fought in and a wound he suffered – 'the wound in my Uncle Toby's groin, which he received at the siege of Namur' – made famous the traditional notion of the hobby horse and its provoking history. It is not incidental here that a hobby horse is at once a self-cure and a consequence of a wound and a war. By definition it organizes and absorbs the attention of its rider, and itself provokes attention. Clearly, for Sterne, the whole notion of the hobby horse raises questions about the paying of attention. And, of course, about our being able to tell the difference between what may and may not be a hobby horse.

In a letter of January 1760, Sterne pursues the divided and dividing question about hobby horses, and so about how a man might be defined by the nature of his attention: 'The ruling passion et *les egarements du Coeur* [and the wanderings of the heart] are the very things which mark and distinguish a man's character; – in which I would as soon leave out a man's head as his hobby-horse.' 'Tell me what attracts or absorbs a person's attention and I will tell you who they are' would be one way of saying this, though not quite Sterne's way (we should note his phrase 'a man's head as his hobby-horse'). Sterne's modern editor Melvyn New glosses Sterne's hobby horse of a man's character with an excerpt from one of his sermons. To understand character, Sterne writes about Herod, we must

> distinguish . . . the principal and ruling passion which leads the character – and separate that from the other parts of it; . . . we often think ourselves inconsistent creatures, when we are the furthest from it, and all the variety of shapes and contradictory appearances we put on, are in truth but so many different attempts to gratify the same governing appetite.

The consistency of character Sterne proposes here is all to do with constant and abiding forms of attention: the same governing appetite (the ruling passions and the albeit overly focused wanderings of the heart). We are not distracted, we just look as though we are, he suggests. There is a pattern even in our inattentions. Vacancies of attention are absorptions elsewhere, as intent and intense as lusts and pastimes: which may themselves be not quite as different as they may seem. We are not divided against ourselves, but far more of a piece than we know. We are not, in any sense, de-centred; but just in states of repeating and repeated displacement. Our attention is only for more of the same. And this raises the question of what fixity of attention – fixating the flow of attention – serves in us.

Sterne suggests we are already filled with attention to what he calls 'the same governing appetite', whether or not it is deemed to be of value. So in what sense can we, or do we, choose, in the telling phrase, to pay attention? Johnson's question in *The History of Rasselas*, like Sterne's question in *Tristram Shandy*, is: to what should we give our attention, and to what do we give our attention (and who decides)? But Sterne's further question is: how seriously should we, and can we, take such questions? *Tristram Shandy*, a parody of sages and a sceptical celebration of the artifices of pleasure, is itself, as Tristram tells us, the story of 'a Cock and a Bull'. 'Lord! said my mother,' the book famously ends: 'what is all this story about? A Cock and a Bull, said Yorick – And one of the best of its kind I ever heard'; the kind being, in Sterne's editor's words, 'a story without direction, rambling, idle, often incredible' (indeed a free association, a story of mobile attention). Sterne is making us wonder which stories – and perhaps particu- larly essentialist stories about character, and attention, and governing appetites – are *not* cock-and-bull stories (i.e. lacking in a certain coherence, plausibility and point). Cock-and-bull stories like *Tristram Shandy* absorb our attention; and Sterne wants us to wonder whether that is the point or the problem. What are we doing when we pay attention to the attention people pay to their hobby horses?

The hobby horse – akin in some ways, to certain psy-choanalytic accounts of so-called perverse states of mind – means absorption with a view to the stopping of time; a form of arrested development or arrested atten-tion; a refuge from the future. And it is, indeed, this link between deprivation and attention (which Johnson refers to as 'vacancies of attention') that Sterne also picks up on. It is as though deprivation itself can stimulate – both force and fix – our attention, and can thus make us attend to our attention. Attention becomes the form and the medium of our desire, prompted by the felt lack of something. We attend to our felt absences.

In Book 6 of *Tristram Shandy*, Sterne returns to the idea of writing and attention; and once again, as with Johnson, it is interestingly the language of fullness and emptiness, appetite and deprivation that he employs. Tristram writes, he tells us,

> one-half full, – and t'other fasting; – or write it all full, – and correct it fasting; – or write it fasting, – and correct it full'. 'When I write full,' he tells us, 'I write free from the cares as well as the terrors of the world. – I count not the number of my scars, – nor does my fancy go forth into dark entries and bye-corners to ante-date my stabs . . . But when . . . I indite [compose] fasting, 'tis a different history. – I pay the world all possible attention and respect, – and have as great a share (whilst it lasts) of that under strapping virtue of discretion as the best of you.

When he is full, he is carefree and unfrightened; when he is fasting, it is – and perhaps he has – 'a different history'; he *does* count his scars, go forth into dark entries and bye-corners to ante-date, i.e. explain, his wounds. And this is what he calls 'paying the world all possible attention and respect'; when he writes full, he fails to do this. Fasting – deprivation – creates attention and respect for the world; and what is then attended to is wounds and cares and terrors. But what is notable about Sterne is that he doesn't privilege either kind of writing; he needs both. It is the combination of the carefree and the terrorized that works for him: 'betwixt both, I write a careless kind of a civil, nonsensical, good-humoured Shandean book, which will do all your hearts good – and all your heads too, – provided you understand it.' What is required in the writing, though, is 'paying the world all possible attention and respect', and also not doing so; the inattention prized alongside the attention. Whereas with the hobby horse there is very little of paying the world all possible attention and respect. In the all too familiar phrase 'paying attention', Sterne reminds us that attention costs us something, and that we are likely and prone in a culture of money to liken attention to money, and so be thinking of investments and returns, profit and loss, gains and drawbacks. By the same token, as it were, we wonder what attention might be like if it was not like paying; paying for something, paying with something; if it was not a kind of investment. If, say, attention was more like affection, or desire, or love. If attention was an experiment in living, rather than a deal or a calculation.

The attentive, terrified, wounded and explaining self is no more real or significant for Sterne than the inattentive, carefree, even careless and fanciful unintimidated self. He wants us to pay attention to both, whatever may be disrespected in the process. One without the other, he suggests, would make for a careworn, uncivil, sensible, humourless un-Shandean book; or a hobby horse. Neither the fasting nor the full self is a diversion or a refuge from each other; they are inextricable and mutually enlivening. And by the same token what we pay attention to can be a retreat or an opportunity, or both.

Attention and its diversions and distractions, Sterne wants to persuade us, are both inextricable and mutually enlivening. But as Thorne reminds us, hobbies and hobby horses are described, conventionally, as what he calls 'divertissement', diversions of the leisured, of the inhabitants of Happy Valleys; and this is, of course, the traditional topos of sacred and secular moralists. There is what we should be attending to, and what we should not be attending to. We know from Sterne that the wound and the hobby horse also somehow go together, as do the full and fasting selves. And we know from Johnson that vacancies of attention are signs of unease. All morality depends upon knowing where and how to pay attention. And to pay attention to morality is also always to pay attention to attention. So, for example, we are familiar in the contemporary language of psychoanalysis with the idea that it is trauma that organizes and narrows – that organizes by narrowing – attention (wounds and hobby horse go together, as do wounds and vacancies of attention). And that morality – like hobbies and hobby horses, and so-called sexual perversions – can be described as, among many other things, a self-cure for trauma, even if that can mean also just the trauma of being desiring creatures.

Johnson says in *The History of Rasselas* that when we are distracted, we are both oppressed and impressed by the tediousness of time. When we fob ourselves off with distractions – when we are just keeping busy – we become bored. Attention nourishes us; distraction depletes us. Attention redeems the time. It makes it feel worth living, and living out. Good attention makes a good life. With its appeal to so-called experience, this has the reassuring clarity that Sterne – in his more con-temporary, and perhaps salutary, way – warns us away from: 'to define is to distrust'. Sterne writes in *Tristram Shandy* that we can tell the difference between attention and distraction – between fullness and fasting – but that this difference only matters because of the combinations that then become possible. To define may be to distrust but it is also to make combination possible. It is not the conversion of (negative) distraction into (positive) attention that we should seek – in Sterne's view – but awareness of the affinities between them. Sterne promotes only the attention that cooperates with inattention. And that is why the difference between them matters.

So when and if you promote *inattention*, what are you promoting? The vacancies of attention in the Happy Valley were, for Rasselas, the sign of need, the precondition for change. The vacancies of attention and the tediousness of time to which they lead inspired his curiosity and his necessary escape from the Happy Valley, in a quest for a knowledge of other people. The preconditions for Tristram's writing – for the writing of a 'careless kind of a civil, nonsensical, good-humoured Shandean book' – are paying the world all possible attention and respect, and also not doing so. Johnson and Sterne, that is to say, in their very different ways are promoting loss of attention, and our paying attention to it. They want us to see the provocations and possibilities in our vacancies of attention.

III

Some great generalization which would finish one's clamor to be educated.

Henry Adams, *The Education of Henry Adams*

I have used *The History of Rasselas* and *Tristram Shandy* as a way of talking about the uses of inattention. But I am not here, in the words of the critic David Bartholomae, 'setting out to write intellectual history or tell a story of influence. I want, rather, to trace a set of common concerns, concerns carried by an odd, deep and persistent vocabulary'. I want to show that a line can be drawn – though not, as I say, necessarily one of direct and acknowledged influence – from Sterne and Johnson through to Freud and then on to some of the essential perplexities of what became British Middle Group psychoanalysis; and particularly Marion Milner's interest in what she calls 'narrow' and 'wide-angled' attention. One way of saying this is that Freud had described psychoanalysis as a new kind of treatment based – for both

the doctor and the patient – on inattention; a treatment
founded on the idea of not, in the traditional way, concentrating. A treatment in which inattention was the
instrument, not the obstacle. The psychoanalyst was a
physician who listened with so-called free-floating (or
'evenly suspended') attention – 'he must give no special,
a priori importance to any aspect of the [patient's] discourse'; and who listened in this way to an unprecedented,
freely associating patient, who was invited to speak without attending to his words; to speak without needing
either to tell a story or to mean what he said; without, in
so far as it was possible, choosing his words, 'selecting
nothing and omitting nothing from what comes into his
mind'. It is as though Freud was saying that only in states
of inattention could certain provocations work; the provocations, that is, of unconscious desire (inattention allowing
us to hear our informing presences). In 'Recommenda-
tions to Physicians Practising Psycho-Analysis' he wrote:

> Just as the patient must relate everything that his self-
> observation can detect, and keep back all the logical
> and affective objections that seek to induce him to make
> a selection from among them, so the doctor must put
> himself in a position to make use of everything he is told
> for the purposes of interpretation and of recognizing the
> concealed unconscious material without substituting a
> censorship of his own for the selection that the patient
> has foregone.

The wrong kind of attention here – the attention that both analyst and patient must suspend – is *censored* attention. Freud is suggesting that attention is primarily, if not essentially, already thoroughly censored (or selective, as we more blandly say). That looking is a way of stopping us seeing; that talking is a way of stopping us speaking; that listening is a way of stopping ourselves hearing. That what we call attending is a process of motivated exclusion; that we concentrate and focus in order to occlude and temper what we might see. That attention evaluates, prohibits and pre-empts; but often unconsciously, without our as yet being aware of it. Psychoanalysis tracks what attention wants and doesn't want to omit. Freud is wondering what we might attend to, and how we might attend, when and if the censorship is lifted. And this means, paradoxically, what we might be attending to if we stopped paying attention.

What we call our attention has been tampered with. So we must avoid, Freud suggests in the same paper,

a danger which is inseparable from the exercise of deliberate attention. For as soon as anyone deliberately concentrates his attention to a certain degree, he begins to select from the material before him; one point will be fixed in his mind with particular clearness and some other will be correspondingly disregarded, and in making this selection he will be following his expectations or inclinations. This, however, is precisely what must not be done. In making the selection, if he follows his expectations, he is in danger of never finding anything but what he already knows; and if he follows his inclinations he will certainly falsify what he may perceive. It must not be forgotten that the things one hears are for the most part things whose meaning is only recognized later on.

This is Freud's version of the Sufi proverb 'Don't learn, listen.' If we deliberately concentrate our attention, all we discover is our expectations and assumptions and presumptions and inclinations; these are what we already know. Our so-called knowledge is all assumption, presumption and preference. Freud offers two prescriptions: suspend deliberate concentration and attention; and allow meaning to take time to emerge. Don't, that is to say, jump to conclusions, or think that you know anything other than your assumptions. A certain kind of inattention leads to the right kind of attentiveness.

When Sterne writes in *Tristram Shandy* that 'to define – is to distrust', he is asking us to consider what or who we should put our trust in, and therefore what we should be attending to. To define, Johnson writes in his albeit definitive dictionary, is 'to circumscribe; to mark the limit; to bound . . . To determine; to decide; to decree.' Attention circumscribed, bounded, determined, decided, decreed, with limits marked, is, of course, the project of Freud's ego as it contends with the concerted disarray of instinctual life, and of the external world. In Freud's story this is what the ego strives, more or less forlornly, to do with and to unconscious desire (and this is what makes it essentially an author of cock-and-bull stories). It does this by contriving apparent vacancies, and narrownesses, of attention. In Freud's story, where there is desire there is always attention and there are always vacancies of attention. Where there is desire, definition does the work of distrust or containment or reassurance. Attention, in the psychoanalytic account, is a compromise, and is therefore compromising.

But the psychoanalyst's problem – and not only the psychoanalyst's – is always to do with the sense in which he already knows what he is looking for. That is to say that it is possible that Freud knows – though he wouldn't, of course, put it like this – that psychoanalysis as an essentialism becomes part of the problem that psychoanalytic method is intending to solve (he believes he knows what man – and to some extent woman – is, and therefore to what we should be paying attention). After all, the psychoanalyst *as psychoanalyst* also has his own expectations and inclinations, whatever his more personal expectations and inclinations might be. The Freudian analyst expects to find primary process thinking, instinctual drive representations, incestuous desire, ambivalence, and conflicts around dependence; he is, as Freud says, 'in danger of never finding anything but what he already knows' (which is why he is so easy to caricature). The analyst, as Freud knows, already knows a lot; and knows a lot about already knowing. How much, given the essentialism he starts from, can his work be, as it were, full of surprises? And if for the most part meaning is only recognized later on, is it recognized once and for all, or in an ongoing and developing way? How much later will it be before meaning emerges? And how will it proliferate? Meaning is only meaning when it is on the move.

The analyst has to stop deliberately concentrating when the patient speaks, and he has to wait an indeterminate time for meaning to emerge; meaning which, by his own theoretical lights, will always be evolving under the aegis of unpredictable, current experiences. What Freud is describing is the difference between the kind of attention paid when we know what we want, what we are looking for; and the kind of attention paid when we are finding out what we want, when we don't know beforehand what we want or what we are looking for, only that we are in a state of wanting and seeking.

Attention can be both instrument and medium. The first kind of attention is intent, determined, and more or less sure of itself (and can be called in psychoanalysis 'a perverse state of mind'); the second kind is more at a loss, and uncertain, tentative and provisional. And so we can read Freud, as he invents psychoanalysis, knowing that he wanted to suspend instrumental attention but without quite being able to bear – or being unable to formulate in the terms available – not really knowing what he might be looking for; not being able to acknowledge desiring without a discernible object of desire. We could describe Freud's project, in other words, as wanting to find a new way of attending to a newish object of attention, instinctual life in language; but also, and more radically, wanting to found a new way of attending to an indefinable object of attention, and with no definitive or defining purpose in mind. There is a Freud who wanted to cure his patients, and a Freud who wanted to enquire about something with them. A Freud who knew both the aim and the objects of his attention, and a Freud who did not. A Freud for whom psychoanalysis was akin to a sexual perversion, and a Freud for whom it was whatever the alternatives to sexual perversion might be. And for all of these Freuds, revisions and redescriptions of attention and inattention were required.

There was a Freud who knew what we should be paying attention to; and a Freud who knew that knowing what you should be paying attention to was a way of not paying attention. Freud as essentialist wants to effectively convert us to what, in his view, man is. Freud the anti-essentialist has – like all anti-essentialists – nothing to convert us to. Except, perhaps, to a suspicion of those who want to convert us. (And those who wish to convert us always know what we should pay attention to; and often exactly how we should be paying that attention.)

So when Marion Milner wrote of narrow and wide attention before she became a psychoanalyst, in her first book, *A Life of One's Own*, she was, in a sense, picking up where Freud left off (whether or not she knew this); or picking up where Freud started. As part of her experiment in finding and making what she calls 'a life of her own', she realized that she needed what she refers to as two kinds of attention, *wide* attention and *narrow* attention (we should perhaps bear in mind here that 'wide' in slang means 'wily' and 'immoral' while 'narrow' also means 'bigoted'). In narrow attention, which, as a 'first way of perceiving', Milner writes,

> seemed to be the automatic one, the kind of attention which my mind gave to everyday affairs when it was left to itself . . . you attend automatically to whatever interests you, whatever seems likely to serve your personal desires; but I could not find anywhere mentioned what seemed to me the most important fact about it, that this kind of attention has a narrow focus, by this means it selects what serves its immediate interests and ignores the rest. As far as I could see it was a 'questing beast' . . . This attitude was probably essential for practical life, so I supposed that from the biological point of view it had to be one which came naturally to the mind.

As a 'questing beast' that from 'the biological point of view' came naturally, narrow attention has, in Milner's Darwinian account, adaptive advantages; it serves 'immediate interests', apparently knowing what these are, but because it ignores everything else, it has a narrow focus (like the expectations and inclinations Freud warns analysts away from); and it is the wider focus, which may or not be in the service of adaptation, that Milner is interested in.

Wide attention, her 'second way of perceiving,' is to do with wanting nothing, and with wanting nothing as the way she discovers that wanting pre-empts experience. This wanting forecloses the discovery of what you might want. These two kinds of attention bring two kinds of provocation: the provocations we know we are looking for, and the unsuspected provocations. (The aim of psychoanalysis, Winnicott once remarked, is to enable the patient to surprise themselves.) In this sense, narrow attention is for essentialists; for the people who, because they know who they are, know what they want. 'It gradually occurred to me,' Milner writes, 'that expectancy might be an obstruction to one's power of seeing . . . particularly active in the sphere of emotion,' whereas wide attention

> seemed to occur when the questing purposes were held in leash. Then, since one wanted nothing, there was no need to select one item to look at rather than another,

so it became possible to look at the whole at once. To attend to something yet want nothing from it, these seemed to be the essentials of the second way of perceiving. I thought that in the ordinary way when we want nothing from any object or situation we ignore it . . . But if by chance we should have discovered the knack of holding wide our attention, then the magic thing happens.

The magic thing that happens is the shock of the new. Wide attention is in the best sense amenable to distraction. 'When at last I did recognize this obstruction to my view,' she writes – the obstruction being the wanting, the knowing what you want, the expectancy – 'then I was able, at least sometimes, to sweep all ideas away from my mind so that immediately real experience, new and indescribable, flooded in.' It is in apparently ridding herself of preconceptions – preconceptions that are an omniscience about wanting – that immediately what she calls 'real experiences' become possible; indescribable because new, and not previously formulated (not yet in language). The questing beast is converted into a kind of mystic through a change of attention. Something is released and comes flooding in; like an orgasm, but unlike because unexpected, unsuspected, uncalculated.

A change of attention is a change of experience. But everything depends – for Johnson and Sterne, as for Freud and Milner – on the relationship between wanting and attention. It is Milner's project to try to disentangle attention from desire and to see where it leaves us. And this frees her to ask the questions that psychoanalysis was previously unable to ask: what happens to attention when we take wanting out of the picture – and what happens to the picture?

Appendix

Greenblatt's Distraction

> I begin to find it no longer in my power to keep my
> attention fixed on things that have little interest for me.
>
> John Ruskin, *Fors Clavigera*

In a review of *The Poetry of Yehuda Amichai* in the *New
York Review of Books*, Stephen Greenblatt wrote that
'Amichai was strangely immune to hate. It was as if hatred
would have distracted him from all that he was determined to observe and to feel.' 'Strangely immune' because
Amichai, a Jew born in Germany who emigrated to Pal-
estine in 1935, the year the Nuremberg Laws were passed,
and who then lived and wrote in Israel – which Greenblatt
refers to as 'his murderously contested part of the world' –
had reasons for hatred; but also because, as Greenblatt
intimates, there may be something uncanny about this
immunity. Hatred can be viral, like some kind of fatal or
contagious illness, and like such illnesses, so utterly pre-
occupying that it can be distracting. There was, in
Greenblatt's characteristically lucid and suggestive

formulation, whatever Amichai was 'determined to observe and to feel' pitted against the distraction of hatred.

That it was 'as if hatred would have distracted [Amichai] from all that he was determined to observe and to feel' is a complex proposition, not least of which is the 'as if' part; as if Amichai could find a feeling so immensely formative as hatred to be a 'distraction' (especially in the light of such a daunting history). By describing it as a distraction, Greenblatt intimates that Amichai was able to be undistracted, or able to distract himself from his distraction; that he in some sense knew what he was determined to observe and to feel, and so knew a distraction when he felt one, knew that the distraction of hatred *was* a distraction (and not a spur, a prompt, an inspiration, a bafflement, or indeed a defeat). Amichai's poetry could have distracted him from his hatred, his hatred could have distracted his poetry, or the poetry and the hatred could have been mutually animating. All of these possibilities might give us pause. We may not be able to imagine lives immune from hatred, but we could imagine how our lives would be different if we could think of hatred as a distraction, and could then wonder what it might be distracting us from; or what might be the preconditions for someone beginning to wonder about hatred in this way, or beginning to wonder in a new way about hatred. That we may not, in any given instance, be simply hating, but using hatred to distract ourselves; that we can link the idea of hatred with the idea of distraction, is itself suggestive. There could be – and Greenblatt's work lets us have a more interesting version of a thought like this – a history of distraction, of distractedness: both of what constitutes it, and of what forms it takes. And all with a view to seeing what describing something as a distraction, in any given period, inhibits, and makes possible (though it is worth noting, in passing, that the meanings of 'distract', 'distraction' and 'distractedness', cited in the *OED*, are virtually unchanging from the fifteenth century to the present day, even if the contexts in which they are used are not).

If culture in the critic Philip Rieff's grand and salutary formulation is 'knowing what to avoid', we gather from Greenblatt's writing that it might also be knowing what a distraction is. There are things we are 'determined to observe and to feel' – and both senses of 'determined' are in play here, just as both senses of 'self-fashioning' are in play elsewhere in Greenblatt's writing – and there is what we are determined to think of as distraction. What would it be not to know what it is to be distracted? Or to put it differently, what kind of mobility of attention do we want, what kinds of attention do we imagine would get us the lives we want?

There is the attention we give, the attention we want to give, and the attention we should give; and the distraction we are subject to. In what Greenblatt calls, and sponsors in his New Historicism, 'an interpretative model that will more adequately account for the unsettling circulation of materials and discourse . . . not outside interpretation, but in the hidden places of negotiation and exchange', we can wonder what kinds of negotiation and what kinds of exchange, both with ourselves and others, our distractedness is (distraction itself being a hiding place). Psychoanalysis encourages us to attend to our distractions as much as we attend to our attention, if not more so (the commitment to free association acknowledges the often disavowed promiscuity of our attention, of desire as distraction; our desire as a continual, and often enlivening, distractedness). And even though much has been said and written about the kinds and forms of attention that we should give to works of art in particular, there is more said, outside of psychoanalysis, about the attention required than about the distractions provoked; or, indeed, about the kinds of distraction, intended or otherwise, that the artefact might want from us, or tempt us with. We might think of any object – person or artefact – demanding certain kinds of attention and attentiveness, and also inspiring certain forms of distraction. Indeed, we might define an object by the distractions to which it leads. Inattention always means attention out of order; and the mobility of our attention is something we are supposed to keep track of (education is knowing what to attend to, and how to attend to it: education puts our attention in order). So one of the things art makes us think about is the qualities of our attention, and what kinds of sabotage our attention is prone to, and drawn to; this has been one of so many remarkable things that Greenblatt's writing has brought to our attention with inspiring and inspired enthusiasm and without portentous gravity.

'Mobility,' he writes, in *Cultural Mobility*,

> can indeed lead to a heightened tolerance of difference and an intensified awareness of the mingled inheritances that constitute even the most tradition-bound cultural stance, but it can also lead to an anxious, defensive, and on occasion violent policing of the boundaries. The crucial first task for scholars is simply to recognize and to track the movements that provoke both intense pleasure and intense anxiety.

Greenblatt is writing here about what he calls 'the fate of culture in an age of global mobility', but I want him to be writing too about the mobility of attention, also partly created by global mobility and the different forms of attention this must conjure. The crucial first task for scholars – and for artists and psychoanalysts, among others – is simply to recognize and to track the movements of attention that provoke both intense pleasure and intense anxiety; and in the full knowledge that anxiety and pleasure can't always be told apart. But in order to do this, you have to know what counts as a distraction; and have something to say about what kind of attention distraction is, both what it may be running away from and what it may be running towards, and why. What

Greenblatt calls 'violent policing of the boundaries' makes vivid and dramatic sense – and especially now when it comes to nation states – but where are the boundaries of our attention, and how do we picture them, or describe them? And what do the boundaries of our attention – boundaries that apparently make the notion of distraction intelligible – have to tell us about the other kinds of boundary that are now paramount? Tracking the movements – of people, of cultural artefacts, of attention – that provoke both intense pleasure and intense anxiety has become one of our abiding contemporary obsessions. But it is also part of the interest of distraction that it can provoke intense pleasure and intense anxiety, often at the same time (think of the adolescent in school daydreaming about sex). And so art becomes, among many other things, one of the best ways we have of thinking about distraction – both through descriptions of distracted states (for example, the 'distracted globe' in *Hamlet*, who himself has 'distraction in's aspect') and also in our wondering what kind of attention art, at any given time, requires of us; how it staves off and incites our distractedness, how it is always in the business of attention-seeking and so must somewhere always be worrying away about our attentiveness (what it is competing with for our attention, which forms of attention give what kinds of pleasure, and so on). After all, if we can be distracted, what does this say about our attention? About its pleasures and its anxieties?

So it is not incidental that it was, indeed, something about distraction that ultimately prompted Greenblatt's influential book *Renaissance Self-Fashioning*, self-fashioning in Greenblatt's account being both the cultural pressures fashioning the self, and a determination not to be distracted from what one is determined to be; and an acute awareness of the force field of distractions that might inform and waylay one's project. In his 1973 book *Sir Walter Ralegh*, Greenblatt describes Ralegh's 'Ocean's Love to Cynthia' as 'another performance . . . in which Ralegh presents himself as a distracted lover cruelly abandoned by his mistress'; a not unfamiliar figure in the poetry of the period, except for the fact that Ralegh was not only distracted because of losing Elizabeth's favour, but also distracted from his love for the queen in his secretly marrying Elizabeth Throckmorton, one distraction leading to another. Greenblatt intimates here that Ralegh was distracted by distraction; that distraction cuts both ways. And that we can never quite know which of the two women was a distraction, and from what. In what Greenblatt calls 'the humanist vision of man freed from any single fixed nature, and able to assume any role . . . the yoking of engagement and detachment', the role of distracted lover begs many questions; not least of which would be the difficulty of telling whether one was distracted from another distraction or from something more essential, or fixed. And this would also, of course, be a theological question.

After writing a dissertation that became the book on Ralegh – 'powerfully struck by the strangeness of Ralegh's long fragment, Ocean's Love to Cynthia . . . and struck by . . . an eerie resemblance between Ralegh's poem, with its tormented sense of a world and self in pieces, and Eliot's Waste Land' – this being struck turned into a larger question: 'Why would a tough-minded Elizabethan courtier, monopolist and adventurer have written poetry at all, let alone poetry that had the ring of modernist experiment?' As though, Greenblatt intimates, the poetry was a distraction, but that the distraction itself might illuminate something about what Ralegh was distracted from; that is, being a tough-minded Elizabethan courtier, monopolist and adventurer. Distraction reflects, and reflects on our attention; and can be of a piece with it. Ralegh, Greenblatt shows, was working out in his poetry his life preoccupations as a newly self-fashioning Renaissance man; what looks like distraction can be attention displaced, or refocused. The poetry is the role-playing life by other means. The book on Ralegh, which was, in a sense, the precursor to *Renaissance Self-Fashioning*, took distraction very seriously. It was, Greenblatt writes, an answer to his question,

> focused on Ralegh's role-playing, his sense of himself as a character in a fiction. Fallen from the queen's favour (as a consequence of secretly marrying), he

cast himself as Orlando Furioso, driven mad by
disappointed love. He went so far as to stage a suicide
attempt . . . and in the same spirit wrote anguished
verses meant to display his state of distraction. (And
distraction, as Eliot the astute student of Renaissance
poetry understood, manifested itself in broken speech,
twisted metaphors, bursts of grief.)

Distraction, here, as casting oneself as something,
behaving not simply as someone else but as a fictional
character. Indeed, in his distraction Ralegh identifies so
powerfully with this fictional character that he stages a
suicide attempt. And of a piece with this, 'in the same
spirit', Greenblatt writes, he 'wrote anguished verses
meant to display his state of distraction' (it was a state
and a performance, a role and a display). The sixteenth-
and seventeenth-century meanings of the verb 'to
distract' given by the *OED* are: 'To draw asunder or
apart; to separate, divide . . . To turn aside, or in another
direction; to divert . . . the attention, the mind, or the
like . . . to draw in different directions; to perplex or
confuse; to cause dissension or disorder . . . To throw
into a state of mind in which one knows not how to
act . . . To derange the intellect, to drive mad.' It is
clearly a suggestive set of definitions; all in their way
also integral to self-fashioning. It was, among other
things, Ralegh's capacity for distraction that struck
Greenblatt; and the kind of language, the poetry,

distraction made possible, 'broken speech, twisted metaphors, bursts of grief'.

As a critic of his time, Greenblatt's account here of Ralegh is haunted by Eliot; the Eliot of *The Waste Land*, the influential critic of Renaissance literature. But also, perhaps, in this context, the later Eliot of *Four Quartets*, who, tracking movements of attention, wrote of '. . . the unattended/Moment, the moment in and out of time,/ The distraction fit, lost in a shaft of sunlight' (*The Dry Salvages*); and also, in *Burnt Norton*, of '. . . a flicker/ Over the strained time-ridden faces/Distracted from distraction by distraction/Filled with fancies and empty of meaning/Tumid apathy with no concentration . . .' 'Distracted from distraction by distraction' measures what Eliot takes to be the modern individual's distance from God, whom one's concentration is for. The only method for writing poetry, Eliot famously pronounced, was to be very intelligent; and he clearly wants us to be intelligent about distraction here (though we should stop talking about intelligent people and just talk about who we really enjoy talking to). But Eliot's question in *Burnt Norton* is: what kind of attention – intelligent or otherwise – can we give to distraction?

Who or what, then, is one supposed to be concentrating on, or what is one's concentration for, what is it in the service of, if this is not preordained? The crucial first task of scholars, as Greenblatt wrote – and of artists, we can add – is simply to recognize and to track the movements that provoke both intense pleasure and intense anxiety; and the pressures informing those movements. And if, as Greenblatt and Catherine Gallagher wrote in *Practicing New Historicism*, 'Analysis of an aesthetic representation must not be a way of containing or closing off . . . complexity but rather of intensifying it', then the attention and the distraction must be included in the analysis as a way of intensifying it, though not as binaries, or contraries, or opposites, because distraction is one of the forms attention takes (raising the questions: is a change of object a change of attention? Is a change of attention a change of object?); it is the attention taken away from supposedly preferred objects of attention, and so a drawing asunder, a diverting if not a dividing of the individual, that threatens confusion, perplexity or even madness. Just as it is Hamlet's distractedness that creates a kind of muted panic in the court, distraction in its early-modern meanings signifies danger. It describes the enigmatic threat of a person who is not concentrating on the right things, a person whose attention has strayed no one knows quite where (distraction creates suspicion in its audience). The radically distracted, like Hamlet, are experienced as opaque – intimidatingly opaque – as though before their distraction they had been well known, reassuringly familiar.

There is then Amichai's in a sense traditional determination not to be distracted (by hatred), which was part of his determination to observe and feel certain things in a certain way, and the poetry this made possible. Distraction as self-sabotage, and destructive of others. And there is Ralegh's determination to be or to perform being distracted; his use of distraction, as it were, as an attempt to redeem himself in the queen's eyes, and recover her favour; and the poetry this may have both enhanced and contributed to. Distraction as facilitation. Eliot's theological diagnosis of distraction, possibly also alluded to by Greenblatt, makes what might be called the traditional case against it – that we need to be distracted from distraction but not by distraction. Distraction is disqualified as an object of desire.

To make the case for distraction can only ever be to make the case for what we need to be distracted from (hatred, say, for Amichai, secular idols for Eliot); making the case for distraction tends to be just another way of privileging particular forms and objects of attention. Distraction always directs our attention back to what our attention is supposed to be directed at (we don't, that is to say, teach distractedness in school, though perhaps we should). And this is what makes Greenblatt's account of Ralegh in his new preface to the 2005 reissue of *Renaissance Self-Fashioning* so telling, alongside his invocation of Eliot. Ralegh can make use of, can perform, can 'display' – as though it were a genre available rather than an essentialism betrayed – what Greenblatt calls his 'state of distraction'. It is a way of fashioning himself with aims in mind: an instance of 'an increased self-consciousness about the fashioning of human identity as a manipulable, artful process' in the sixteenth century. Like an early (Jamesian) pragmatist, Ralegh is seeing where his distraction gets him; he is really trying out distractedness, trying it on, like an actor with a new part, but to an extreme.

'At a deep level,' Greenblatt writes in *Hamlet in Purgatory*, 'there is something magnificently opportunistic, appropriative, absorptive, even cannibalistic about Shakespeare's art.' This is an art – and an imagination – that is voracious, not fastidious. It is also, perhaps unsurprisingly, the description of an actor. And an actor in a play – unless he is acting distractedness – can only be distracted from his part. What, in any given instance, we are going to call distraction, then, will depend on certain sureties about where our attention should be. Whether as evasion or protest, when it comes to distraction we are sometimes more certain – that is, more informed – about what we should be paying attention to than about where our attention has gone (attention unmoored is the threat; as though the self – the self as representation – effectively comes into being through organizing itself around objects of attention. When Proust's Marcel famously wakes up disorientated in the bedroom at Combray, he literally reconstitutes himself through attending to the objects in the room). And when it comes to distraction, as it often does, we may be less sure about where our real enjoyment is. Unless, that is, we can make distractedness a part of our repertoire, unless we can find ways of describing it as something we do, with interesting objects in mind, with a view to fashioning an identity. And this involves, among other things, being able to tell the distraction from the other thing. Because distraction is another form of attention, we may not always be able to tell which the distracted state of mind is. The authorities can tell us.

So when Greenblatt writes, in *Renaissance Self-Fashioning*, about what he calls Sir Thomas More's 'distracting engagements' with civil work and family, which took him away from his real preoccupations, he glosses this by saying: 'There is always, it seems, a "real" self – humanistic scholar or monk – buried or neglected, and More's nature is such that one suspects that, had he pursued wholeheartedly one of these other identities, he would have continued to feel the same way.' More's distracting engagements imply what Greenblatt calls 'a secret reserve, a sense of life elsewhere, unrealized in public performance'; and yet the whole notion of distracting engagements may lead, as he makes clear, not to possibilities of restoration and nourishment, but to suspicions about a 'real self' – an undistracted self – and what such a representation, such a figure, might be doing for More (the fantasy of an undistracted self may sponsor a capacity for distraction, just as fantasies of uninhibitedness keep us inhibited: that's what they are for). The figure of the undistracted self might undo something it is attempting to consolidate; it may be a consolation that More fobs himself off with. 'For his life seems nothing less than this,' Greenblatt writes, and less can't help but make us think of More, 'the invention of a disturbingly unfamiliar form of consciousness, tense, ironic, witty, poised between engagement and detachment, and above all, fully aware of its own status as an invention.'

'Poised between engagement and detachment' suggests the uses of distraction as part of the invention: you can only be engaged and detached if you can distract yourself. And this might require not abrogating the idea of a real self, but including it as one of the parts in an ongoing drama of survival. 'One consequence of life lived as histrionic improvisation,' Greenblatt writes of More – and he could be writing of Ralegh – 'is the category of the real merges with that of the fictive.' The double act of the distracted self and the real self keeps the show on the road. But this breeds something darker, in Greenblatt's account: 'There is behind these shadowy selves,' he writes, 'still another, darker shadow: the dream of a cancellation of identity itself, an end to all improvisation, an escape from narrative.' The strain of this 'protean adaptation', the work of self-invention, could be too much. At this moment what is being considered is that there might be literally nothing to be distracted from, except, perhaps, the strain of protean adaptation; or that there is merely a series of distraction, but nothing one is distracted from: protean adaptation with no pleasure in what is there to be adapted to. Indeed, protean adaptation might feel like endless, futile distraction. If distraction is a form of adaptation, we can still wonder, as Freud did, what adaptation itself might be a distraction from. So the art in a distracted life may need to be artful about distraction. If, as Greenblatt writes of More, 'the dialectic of engagement and detachment is among those forces that generated the intense individuality that, since Burckhardt, has been recognized as one of the legacies of the Renaissance', then distraction, which is often at once engagement and detachment, has a part to play. The art in a distracted life may need to be artful about distraction in part because it can sometimes be so difficult to tell the difference between engagement and detachment, as each makes the other possible. Psychoanalysis, one might say, was a modern attempt to tell this difference; and to find out what else one might say in the telling.

Acknowledgements

These essays began as talks and lectures, and have been revised for this book. 'Vacancies of Attention' was given at the Provoking Attention Conference at Brown University, Rhode Island, in 2017, organized by Amanda Anderson and David Russell. 'Shame and Attention' was published in *Salmagundi*; I am grateful as ever for the editing and hospitality of Bob and Peg Boyers. 'Greenblatt's Distraction' was given on the occasion of Stephen Greenblatt receiving the Holberg Prize, in Bergen, in 2016. The three essays on attention were also given, in different versions, at the University of York.

I am grateful, as always, for the friendship and conversation of Hugh Haughton, Kit Fan, Matthew Bevis, Chris Oakley, David Russell, Stephen Greenblatt, Ramie Targoff, Barbara Taylor, Norma Clarke, Tom Weaver, Brian Cummins, Geoffrey Weaver and John Gray. Simon Prosser is my reliably inspiring editor. And Judith Clark, as always, has made this book and much else possible.

Permissions

Permissions

Eric Griffiths, *If Not Critical* (Oxford University Press USA, 2018). Reproduced with permission of the licensor through PLSclear.

Bernard Williams, *Shame and Necessity* © 1993. Republished with permission of University of California Press; permission conveyed through Copyright Clearance Center, Inc.

Marion Milner, *A Life of One's Own* (1934). Republished with permission of Taylor & Francis; permission conveyed through Copyright Clearance Center, Inc.

Stephen Greenblatt, *Cultural Mobility: A Manifesto* (Cambridge University Press, 2009). Reproduced with permission of The Licensor through PLSclear.

Stephen Greenblatt, *Sir Walter Ralegh: The Renaissance Man and His Role* (Yale University Press, 1973). Reprinted by kind permission of the author.